Titles in This Series

3 **L. E. Sadovskiĭ and A. L. Sadovskiĭ,** Mathematics and sports, 1993
2 **Yu. A. Shashkin,** Fixed points, 1991
1 **V. M. Tikhomirov,** Stories about maxima and minima, 1990

Mathematical World • Volume 3

Mathematics and Sports

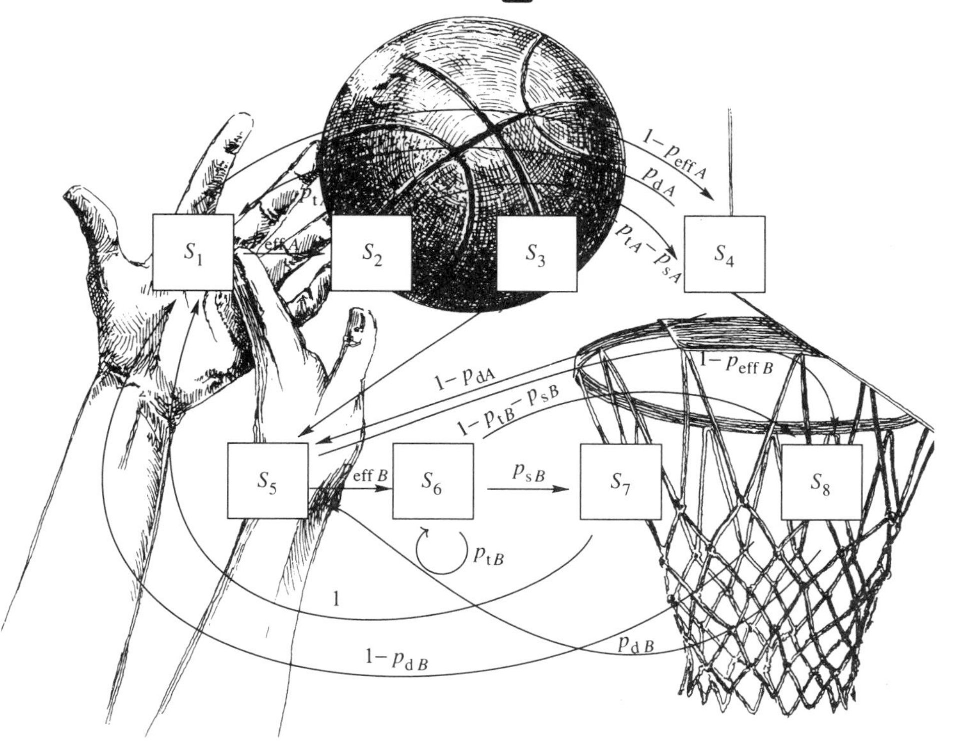

L. E. Sadovskiĭ and A. L. Sadovskiĭ

Translated from the Russian by

S. Makar-Limanov

American Mathematical Society

Л. Е. САДОВСКИЙ
А. Л. САДОВСКИЙ

МАТЕМАТИКА И СПОРТ

«НАУКА», Москва, 1985

Translated from the Russian by S. Makar-Limanov
Translation edited by Simeon Ivanov

2000 *Mathematics Subject Classification.* Primary 91F99;
Secondary 91B99, 90C05, 91A80.

Library of Congress Cataloging-in-Publication Data

Sadovskiĭ, Leonid Efimovich.
[Matematika i sport. English]
Mathematics and sports/L. E. Sadovskiĭ, A. L. Sadovskiĭ; translated from the Russian by S. Makar-Limanov.
p. cm.—(Mathematical world; v. 3)
Translation of: Matematika i sport.
Includes bibliographical references.
ISBN 0-8218-9500-1 (alk. paper)
1. Sports—Mathematics. I. Sadovskiĭ, A. L. (Alekseĭ Leonidovich) II. Title. III. Series.
GV706.8.S2313 1993 93-23024
796′.02–dc20 CIP

♾ The paper used in this book is acid-free and falls within the guidelines
established to ensure permanence and durability.

This publication was typeset using AMS-TEX,
the American Mathematical Society's TEX macro system.
Visit the AMS home page at URL: http://www.ams.org/

10 9 8 7 6 5 4 05 04 03 02 01 00

Table of Contents

Preface ix

Chapter 1. Mathematics and Sports (in Place of a Foreword) 1

Chapter 2. What is Applied Mathematics? 5

Section 2.1 What are the distinctive features of applied mathematics? 5

Section 2.2 Mathematical models 8

Section 2.3 Operations research 9

Chapter 3. Why Five Sets? (Mathematical Modeling of Tennis) 13

Section 3.1 A little history 13

Section 3.2 The arithmetic of tennis 14

Section 3.3 How long should one play? 15

Section 3.4 Elementary concepts of probability theory . . 16

Section 3.5 The model of a game is a Markov chain 18

Section 3.6 Play! 19

Section 3.7 Finishing the game 20

Section 3.8 Vector operations 21

Section 3.9 Now let us go on to complete the set 22

Section 3.10 Let's have it out 24

Section 3.11 Practice makes perfect 25

Section 3.12 Model of a tiebreaker 26

Section 3.13 Markov chains and basketball 27

Chapter 4. Those Judges! 35

Section 4.1 What is an examination by experts? 35

Section 4.2 Rankings 36

Section 4.3 Shortcomings of the majority principle 37

Section 4.4 Judging figure skating 40

Section 4.5 Multiround examinations by experts and their modeling 45

Section 4.6 Hierarchic examination (judging competitive gymnastics) 47

Section 4.7 Examination by experts. An overview 49

Chapter 5. Records! Records! 53

Section 5.1 Random variables 53

Section 5.2 Faster! 56

Section 5.3 Higher! 58

Section 5.4 Stronger! 64

Section 5.5 A jump into the twenty-first century? 67

Chapter 6. Linear Programming and Sports **71**

Section 6.1 Positioning players on a basketball team **71**

Section 6.2 Soccer clubs and players **75**

Section 6.3 Some basic concepts and facts **78**

Section 6.4 The problem of an athlete's diet **84**

Section 6.5 Linear programming problems **86**

Section 6.6 Corner points and convex combinations **90**

Section 6.7 Corner points and feasible solutions **93**

Section 6.8 The simplex method **95**

Section 6.9 Indoors or out? **96**

Section 6.10 Some general inferences **99**

Chapter 7. Game Models **103**

Section 7.1 Meteors vs. Pennants (on a soccer theme) .. **103**

Section 7.2 Matrix games **107**

Section 7.3 Problem of the final spurt **114**

Section 7.4 Games against nature **118**

Section 7.5 How to wax skis **118**

Section 7.6 The iron game **121**

Section 7.7 Matrix games and linear programming **126**

Section 7.8 Zany Zebras at Mayapple Leafs (on a hockey theme) **127**

Section 7.9 How to form a swimming team **129**

Chapter 8. Organizing Competitions is an Operations Planning Problem 131

Section 8.1 The Olympic system 131

Section 8.2 Round robin 132

Section 8.3 The Scheveningen system and Latin squares . 134

Chapter 9. Classifications in Sports 139

Section 9.1 Classification principles 139

Section 9.2 The international tennis classification 144

Section 9.3 Domestic tennis ranking systems 145

Chapter 10. Conclusion 147

Section 10.1 Do not get carried away 147

Section 10.2 Inducements to research 148

Section 10.3 A brief survey of applications 148

Section 10.4 The real conclusion 150

References 151

Preface

This book was written seven years ago in the former Soviet Union for high school and college students. At that time some of the most popular sports in the USA—such as football, baseball, and golf—were practically unknown over there. Now there is a football team, the Moscow Bears, formed by students of the Moscow State University. Baseball and golf have also become popular.

In this volume you will find some examples of possible mathematical applications in tennis, swimming, track and field, and so on.

I would like to express my thanks to the translator and to the editor for their helpful remarks and for their time and attention.

Alex Sadovskiĭ
April 1992
Kingsville, Texas

1

Mathematics and Sports (in Place of a Foreword)

It may seem, at first sight, that mathematics and sports are very far apart. Indeed, many young people mistakenly believe that learning math is one thing and going in for sports is quite another. They think so because they are inexperienced, and perhaps because their school focuses on the exact sciences and neglects physical education. Admittedly, too many bright students look down on games and physical training. At the same time, though, many scientists, including mathematicians and physicists of the older generation, take the sports they go in for very seriously, knowing, as they do, that sports promote a person's all-around development, intellectual as well as physical.

In recent decades dramatic changes have occurred in life and in the quality of education, especially in the area of exact sciences. The greater flow of information has increased psychological stress at work and at school. The new conditions of life, study, and work make it imperative for the young—and not-so-young—to have mental and physical stability. We are convinced, from observation and our own experience, that those working in mathematics and physics are particularly in need of such stability. Creative scientists know the joy of discovery (an extraordinary experience not everyone is familiar with), as well as the fatigue coming on the heels of extreme mental strain. Norbert Wiener wrote: Severe work of a research nature drains one dry, and without an ample opportunity to rest as intensely as one has worked the quality of one's research must go down and down.[1]

People seek mental relaxation in a variety of ways. Some play bridge (they call it a mathematician's game), others prefer chess (the very thing for a highbrow), while others still would go on a hike now and then—but very few turn to sports. Most people believe that neither bridge nor chess, nor the Japanese game of Go, nor any other game requiring great intellectual concentration is truly relaxing. One current—perhaps not indisputable—opinion is that it is impossible to engage seriously in mathematics and chess at the same time. They cite especially the example of Emanuel Lasker, who, upon becoming an

[1]Norbert Wiener, *I am a mathematician*, M.I.T. Press, Cambridge, MA, 1970, p. 127.

outstanding chess player and world champion, quit mathematics, in which he had been active and obtained results that went into textbooks.

Some scientists claim that strong tobacco and spirits clear the head and spur creativity. It would be well, however, to try other means: to exercise, jog, swim, or learn to play games like tennis, basketball, badminton, volleyball, and so on. We shall cite again Norbert Wiener, who found that his writing could be done best when his work alternated with simple pleasures like walking or swimming, which involved no intellectual effort. Those who prefer intellectual games should know that in sports and games one's mind, education, and foresight are far from unimportant. A really good tennis player, for example, must work at his technique no less perhaps than a violinist has to work at his. But out on the court, he usually has to deal with an equally strong opponent. And here, it is tactics, gumption, and foresight that carry the the day. It is not fortuitous that a great majority of good tennis players are well-educated, smart men and women or that tennis should be so popular among academics. Tennis, however, is not an exception, similar things being true for other sports and games. According to authoritative opinion, modern sports have been getting more and more intellectualized. Another thing to be borne in mind is that not only checkers, chess, cards, or billiards are a source of interesting problems; other sports provide them as well. Mathematical methods are increasingly applied in sports. Just think how many yet-unsolved problems arise when we study the interaction between ball and racket or between ball and court. Among other things in this book you will find tennis discussed in terms of Markov process theory, the system of judging adopted in figure skating described in terms of examination by experts, and so on.

Of course, mathematical statistics is applied to estimate an athlete's chances of success, identify the best training conditions for him or her and the measure of its effectiveness, and to process the readings of load-controlling sensors. Information theory makes it possible to estimate the amount of eyestrain in mountain skiing, table tennis, and so on. Mathematics and physics help identify the best shape of rowboats and oars. Chebyshev's ideas on cloth cutting are believed to have been applied in constructing the tennis ball cloth cover.[2]

At the same time, sports have a beneficial effect on one's intellectual activities, state of mind, and will power. Scientists who swim, play tennis, run, ski, and go mountain climbing are well aware of it.

Admittedly, the extraordinary creative longevity of many of our outstanding mathematicians and physicists is due to their affinity for sports.

To be named among noted scientists keenly interested in sports are B. Pontecorvo, J. Littlewood, and R. Peli. Niels and Harald Bohr were outstanding soccer players when at the university. Niels was an excellent skier. Albert

---

[2]Professor Ventsel, of the Moscow College of Transport Engineers Applied Mathematics Department, related that in her student days at Leningrad University, her teachers used to tell of Chebyshev's public lecture on cutting. Tailors attended in force. His opening words were: "Assume that man is shaped like a sphere."

Einstein liked to play the violin, of course, but he also was an enthusiastic yachtsman. The mathematicians and physicists who go in for mountain climbing merit a separate book. We can also cite Charlie Chaplin, who wrote that whenever he was badly upset, he would pick up his tennis racket and, going to the practice wall, would strike the ball against it until he calmed down.

If we compare children who have received physical education with those who never cared for sports, we shall see that the former cope with problems better and have more stamina.

There are many paths leading to amateur sports, to a way of life in which study and, later on, work and creative endeavor go hand in hand with exercise. It is never too late the follow these paths.

A great deal is done by the schools which offer physical training, skiing, swimming lessons, and preparation for the standard fitness tests. Besides, public education departments, Young Pioneer Palaces, and sports societies sponsor numerous children's athletic schools welcoming especially grade schoolers.

In the colleges, gym is supervised by physical education and sports departments. Besides—or instead of—the required courses, undergraduates may join sections dedicated to swimming, wrestling, gymnastics, track and field, weight lifting, basketball, volleyball, and so on. The more enthusiastic and perseverant become rated athletes. This does not hinder but, on the contrary, promotes studies. All one needs to do is manage the time sensibly. With all the sports societies that have their sections everywhere, opportunity is not lacking. Therefore, on our path to sports, a great deal depends on ourselves. Sports build up self-reliance and bring success in one's service to society.

It may be that the readers of this book believe in sports and find time for them. Then again, they may range from those familiar with the exact sciences and ignorant of sports, to those familiar with sports and ignorant of the exact sciences—mathematics in the first place.

Of course, sports provide an inexhaustible source of fascinating and challenging problems in medicine, biomechanics, hydro- and aerodynamics, social science, statistics, etc. These problems are examined, solved, and described by experts in each respective field. It would be too much to try to cover it all in a publication of this size; and it would be too much for us, too. In this book, we shall try to call the readers' attention to the fact that it is possible to study many situations in sports from a mathematical perspective and that it is desirable to obtain more valid quantitative and qualitative estimates of the things happening in sports. We shall consider only some of the situations that can be investigated by means of operations research, a new field that developed over the recent four decades and has gained prominence in applied mathematics.

2

What Is Applied Mathematics?

2.1. What are the distinctive features of applied mathematics? By pure and applied mathematics, we shall mean the "academic" mathematics, taught by university mathematics departments and based entirely on the deductive method, and mathematics in the form acquired by it in the course of its application.

Naturally enough, many concepts, statements, and methods, whether in pure or applied mathematics, have the same—or almost the same—meaning (e.g., the Pythagorean theorem). Now, however, we shall focus on distinctions.

Existence of a solution. The question, "Does this problem have a solution?" is not as simple as it may appear at first sight; often enough a "pure" or an "applied" mathematician may give diametrically opposite answers to it. Leaving the intricate philosophical and logical details aside, let us illustrate this by an example.

There is a popular game among American undergraduates, called Hex. It is a game for two, played with black and striped pieces on a four-sided board comprised of hexagons (a tile floor will do perfectly). The board usually has eleven hexagons on each edge (see Figure 1 on the following page). Two opposite edges of the board are designated "black", and the other two sides, "striped". Each player has a supply of pieces of one color. The players alternately place a piece on any vacant hexagon. The object of the game is for each player to complete an unbroken chain of his pieces between his sides of the board.

The question naturally arises: Is there a winning strategy for the first or second player? (A strategy consists of indicating the move in each situation already arrived at). This question was answered by the noted American mathematician John Nash. He proved that a winning strategy exists for the player who starts the game, but not for the other. The following is an outline of his ingenious proof.

First of all, it can be proved (we leave this to the reader) that one of the players must win the game—it cannot end in a draw. This said, we now prove

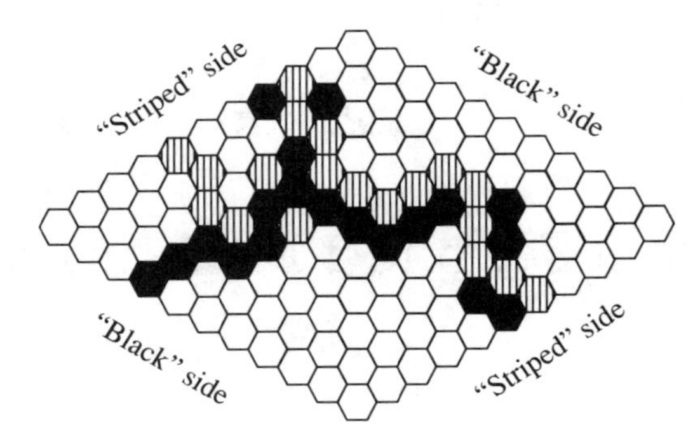

FIGURE 1

the existence of a winning strategy for the first player by contradiction, i.e., by assuming that no such strategy exists. This means that whatever the first player may do, the second player can still avoid being defeated; i.e., he will win because there can be no draw. But then, the second player has a winning strategy (Why, do you think?).

Let the first player proceed as follows. He puts the first piece on any hexagon and then, paying no attention to it, makes his moves, taking advantage of the second player's winning strategy (as if he were the second player). He continues in this way until he, by dint of this strategy, comes to need the place already occupied by the first piece. At this point he puts the next piece on any vacant spot and goes on playing, disregarding it, taking advantage of the second player's winning strategy, and so on. Thus, after each move, the first player has the position assigned by the strategy for the second player, and one covered hexagon besides. Hence, his opponent cannot finish the game, no matter what is his next move. And as the game cannot end in a draw, the first player must win, contrary to the assumption. The contradiction obtained proves the existence of a winning strategy for the first player; it is clear, therefore, that there can be no such strategy for the second player.

It might seem that the winning strategy problem is resolved. But here comes a player, who asks the mathematician: "How should I play to be sure to win?" Analysis of the foregoing proof warrants only this answer: "Check all the possible strategies (there is a finite number of them); by dint of what has been said, at least one of them is a winning strategy, and all you have to do is follow it." "Check all the possible strategies? Have you considered their number?" "Now, that has nothing to do with mathematics," the mathematician will reply. "Ask the engineers to build a checking device. I did my part!"

That was a "pure" mathematician. An "applied" mathematician, on the other hand, would construct a solution taking into account the real circumstances of the case. It is easy to verify that the total number S of all possible strategies (at least for the first player) satisfies the inequality

$$S > 121 \cdot 119^{120} \cdot 117^{118} \cdot 115^{116} \cdot \dots \cdot 101^{102}$$

(recall that the playing field consists of 121 hexagons).

The right part of this inequality certainly exceeds $100^1 \cdot 100^{120} \cdot 100^{118} \cdot \dots \cdot 100^{102} = 10^{1111}$. No device will ever be able to exhaust such a number of alternatives.

Thus, we have before us the statement that the problem has a solution, which is quite legitimate in terms of "orthodox" pure mathematics but unacceptable in terms of applied mathematics. Roughly speaking, the divergence of the two approaches is due to the fact that the finite—in "orthodox" terms—total number of strategies has proved in practice to be ... infinite. Therefore, although an abstract solution of the problem does exist (it has been proved that there is a winning strategy for the first player), the "applied" mathematician will say that the Hex problem has no solution (it is impossible, in the general case, to indicate a realizable algorithm for finding this winning strategy).

Method of reasoning. "Not quite" is a notion that does not exist in pure mathematics. There, whatever is "not quite exactly defined", is, in fact, not defined; and whatever is "not quite rigorously proved", is not proved. When solving a problem in pure mathematics, one can go from one statement to another, proceeding from the conditions of the problem, only based on strict rules of logic.

Not so in applied mathematics. Certainly, deductive reasoning plays a highly important part in it too. Still, there are here equally important arguments of a different kind, called "heuristic", "probable", "rational", and so on. Such arguments are unacceptable from the standpoint of pure mathematics. Nevertheless, if intelligently applied, they lead to correct practical results. Such arguments are typical of all disciplines (physics, chemistry, biology, medicine, etc.), except pure mathematics, so that in this respect applied mathematics lies, as it were, in the interface between mathematics and these disciplines.

Heuristic reasoning may include analogies, numerical and physical experiments, general deductions based on the analysis of typical cases (called "incomplete induction"), and similar other arguments. None of these methods are considered valid in pure mathematics. Even so, they are quite legitimate as far as applied problems are concerned and are regularly employed.

In applied mathematical reasoning, mathematical concepts usually stand for real objects. This is why information following from the "physical sense" of an applied problem, though not explicitly contained in the statement of the problem, often proves useful.

In a number of cases, however, the deductive method turns out to be the most expedient, if not the only possible one. So applied mathematics employs all methods of reasoning.

Then why cannot all constructions be carried out as rigorously in applied mathematics as in pure mathematics? The point is that it is often possible to obtain solutions heuristically where purely deductive methods fail or require a stupendous effort, not worth undertaking. Apart from that, transition from a real object to its mathematical model (see below) is always heuristic and is accurate only to a certain extent. So the solution of the mathematical problem is merely a part of the complete investigation.

2.2. Mathematical models. An expert in applied mathematics has to deal with mathematical models all the time. Models may be geometric figures, number sets, equations, systems of equations, etc., that describe the properties of the real object or phenomenon under study.

Consider the following simple example. Suppose we would like to know how much liquid can go into a glass. We can find that, for instance, by filling the glass with water and pouring it out into a graduated contained. And now we say that the glass in a round cylinder, with a base diameter d and a height h.

By doing so, we convert to a mathematical model that allows us to obtain the answer, $h\pi d^2/4$, without trial but also without considering the imperfection of the actual shape of the glass, surface tension, and so on.

Certainly, a mathematical model describes the real object only approximately. Sometimes, however, a mathematical model may describe it in an entirely wrong way, or, in other words, be inadequate to the real object. Constructing a mathematical model is a highly responsible matter. There may be many different, nonequivalent models of a real object. The search for an adequate and, at the same time, sufficiently simple model is often dramatic. Furthermore, while studying a model, we may run into unexpected mathematical difficulties.

Every scientific approach involves construction of models of the phenomena under study. A model is, in a sense, simpler than the object itself; it does not usually simulate all the features of the object, but only those of most importance to the investigation, so that it can be studied more conveniently. It all depends on the object (phenomenon, situation) under study and on the characteristics the model takes into account. Thus, a system of linear algebraic equations (e.g., the Kirchhoff law, familiar from high school physics) is a mathematical model of currents flowing through an electrical circuit. An automatic train driver (or an autopilot) is a physical model reproducing the work of a train driver (or a pilot). A system of elastic rods may serve as a physical model of a building structure, and a system of equations relating the tensions in these rods may be its mathematical model. N. Wiener and his collaborators have constructed a mechanical, and then a mathematical, model simulating the shaking of the hand of a person with impaired coordination of movements.

Nor is there anything unusual in modeling. Using the means and the language of mathematics in industrial, economic, or any other design is, in effect, mathematical modeling. One and the same object (phenomenon, situation) may have several nonequivalent models, depending on the characteristics each model reflects. However, even when the same characteristics are reflected, the models may differ a little.

A mathematical model must be adequate to the real object under study;

i.e., it must correctly describe the object based on certain characteristics. This is a matter of primary importance. A model that is adequate with reference to one system of characteristics may be inadequate with reference to another. For example, in §3 we construct a mathematical model of tennis, adequate to the game by its basic characteristic of the scoring procedure in a game (set). This model, however, does not include the emotional or psychological factors or a player's adaptation to his opponent's style of playing. Then the model is refined, and one more characteristic is introduced, that of a player adapting one-self or learning in the course of the game. This model, however, is still inadequate to the real process with reference to some other features.

A model should be relatively simple. The existing methods (computers) should make it possible to analyze the model with reference to the characteristics selected. Usually, the more adequate a model is to the real process, the more complex it gets. Thus, the requirements of simplicity and adequacy are, in a way, opposed to each other.

In a first approximation, characteristics involved in modeling fall into two categories. One of them includes variables amenable to sufficiently accurate measurement and control; these are controlled variables. The other category is comprised of stochastic variables, which are of a random nature and cannot be measured accurately. The aforementioned model of tennis has stochastic characteristics and is described in terms of the theory of probabilities and random processes. Also stochastic is the model of forecasting competition results (§5). On the other hand, the model of role distribution in team sports (basketball, hockey, etc.) is a deterministic one (§6).

Mathematical models, the purpose of which is to show why one or another of the possible decisions should be chosen in a given situation, are studied in operations research, a very important division of applied mathematics.

2.3. Operations research. The need for making decisions is as old as man himself. Decision problems begin in the cradle and keep emerging throughout life, arising in such areas as education, daily living, one's job, getting about, etc. Society, its organizations, and its governing bodies all have to make decisions on a myriad of problems.

There are problems of efficient control of the economy, industrial association, and individual enterprise; problems of efficient control of resources, transportation, and freight traffic; problems of queuing at clinics, hairdresser's, service stations, and so on. Solutions of these problems are generally concerned with maximizing income, minimizing costs, minimizing service time, maximizing speed, and so on.

Decisions have to be made in numerous situations in sports: in arranging practice sessions and competitions, recruiting players, assigning positions in the team, choosing the tactics, and so on.

In point of fact, everyone has to make decisions almost every minute of the day on what to wear, when to do what, where to eat, and by which route to go to work, to the theater, or on a visit.

Animals, too, must make decisions when searching for food, defending themselves from a predator, rearing their young, and in other situations. However, the ability to make *reasoned* decisions must be one of the things that distinguish man from the rest of the animal world. Still, we can say that someone is "engaged in operations research" only when applying mathematical methods, employing mathematical means. In reality, decisions are usually made intuitively (e.g., in self-defense), based on experience (e.g., in selecting clothes or meals), by comparing (not always exhaustively or indisputably) various alternatives, or in other nonmathematical ways.

On the other hand, many situations (including some of those mentioned above) are so involved, and the outcomes of the decisions may prove so significant as to make a preliminary quantitative and qualitative analysis mandatory. In these situations scientific methods—mathematical method above all—has to be applied.

" 'Measure twice before you cut once', says the proverb. Operations research is just such a mathematical 'measuring' of prospective decisions, helping to save time, effort, and resources, and making it possible to avoid errors too costly to learn from. The bigger, the more complex and expensive the project, the less it should depend on arbitrary decisions, and the more important the role of the scientific method, which makes it possible to evaluate in advance the outcomes of each decision, rejecting the unfeasible alternatives, and recommending the most preferable ones" **[2]**.

Operations research began as a scientific discipline in the early years of World War II when it become necessary to provide recommendations on military problems such as transportation of troops to the theaters of war, efficient use of weapons, disposition of means of destruction, and so on. Subsequently, operations research was brought to bear on a diversity of nonmilitary problems, such as allocation of resources, inventory control, organization of production, transportation, and numerous other problems.

Solving a problem by operations-research method consists of a number of stages of which we shall note the following (they are fairly closely related).

1. Formulating the problem, situation, process; giving a verbal description of the object of investigation; and interpreting them.
2. Choosing the method of investigation and constructing a mathematical model (one of those possible) of the problem.
3. Studying the analyzing the model; formalizing the relations (identifying the mathematical relations) between the parameters or characteristics of the model.
4. Constructing criteria (objective functions) for evaluating possible solutions.
5. Optimizing the criteria by applying them to numerous possible (feasible) solutions, i.e., choosing optimal solutions (preferred for certain reasons).
6. Comparing the model and the solutions with the real object.
7. Correcting the mathematical model or constructing another, based on a different method of mathematical formalization.
8. Issuing recommendations.

9. Implementing one of the optimal solutions; note that this final stage lies beyond the boundaries of operations-research theory, being carried out by a person (a group, an administrative body, etc.) responsible for making the decision. All the preceding stages are to facilitate and justify this final stage.

3

Why Five Sets? (Mathematical Modeling of Tennis)

3.1. A little history. Over a hundred years ago, in February 1874, W. Wingfield, a retired British major, patented a new game. He called it tennis. In fact though, tennis has a very long history. According to some records, it existed in ancient Egypt. More recently, in the thirteenth century, it was played in France where the players threw the ball to each other with the palms of their hands. It was called royal tennis or "game of the palm" ("jeu de paume"). There was a popular game for two in England during the thirteenth through eighteenth centuries. One player hit the ball with his palm against the wall; when the ball bounced off, it was struck by the second player, and they went on like this until one of them made an error.

Tennis was first described in Russian Literature by Leo Tolstoy in his novel *Anna Karenina*. Evidently, the description refers to the mid-80s of the past century: "The players, separating into two parties, took their places on the carefully levelled and tamped croquet ground, on either side of a taut net... Sviyazhskiĭ and Vronskiĭ were both playing very ably and seriously. They watchfully followed the ball being thrown to them, nimbly, neither rushing nor tarrying, ran up, paused before the jump, and, hitting the ball precisely and confidently with the racket, threw it over the net" (Part 6, Chapter 22).

Tennis in the times of Anna Karenina was not much like the modern game. It was played regardless of the ground surface and was a kind of entertainment involving some physical effort. The clothes worn for the game, especially ladies' long dresses, were inconvenient, the rackets were loosely strung, the strings were thick, and the balls had no cloth covering. Incidentally, in bygone times balls were made of pieces of cloth stuffed with horsehair. About five centuries ago, Louis XI ordered balls to be covered in leather and wool. Subsequently balls were made of raw rubber. For some decades now tennis balls have been made of rubber covered with felt.

A great role in the spread of tennis in Russian was played by the creation of clubs. The St. Petersburg Cricket and Lawn Tennis Club was founded in 1860. The All-Russia Lawn Tennis Club Union, which united forty-seven clubs, was established in 1908. The beginnings of tennis as a competitive sport may be dated back to the 1890s, when new rules of the game were established. In 1891, E. M. Dement′ev, M.D., published the first book on tennis in Russian.

Tennis quickly became popular among young people and gym students. It was due, in part, to the fact that the favorite game at that time was Russian lapta.[3] At some schools, lapta was one of the physical education activities. They also played a game in which a ball was hit back and forth over a high net with small wooden paddles.

In 1913, at the proposal of B. A. Ulyanov, Secretary of the Lawn Tennis Union, Russian terms corresponding to ad in, ad out, deuce, etc., were adopted.

Russian tennis players took part for the first time in an international tournament at Stockholm in 1903. The first international tennis tournament in Russia was held in 1913. Only a small number of people played tennis at that time: there were 154 rated players in 1912, and 203 in 1913.

Once exclusive, tennis became a popular sport. In Moscow alone, there are about 1,500 rated players and some tens of thousands of formerly rated players and others who play tennis on a regular basis. Let us note that tennis, currently played by 120 million people in 193 countries (as opposed to 40 million football (=soccer) players), is one of the most healthful and exhilarating sports. Tennis is an unequaled antidote to the bane of these times—the sedentary way of life. In the rate of growth and popularity, it has recently outstripped all other sports. Besides suiting people of all occupations, tennis can also provide immense moral and physical satisfaction. Indeed, it is said that in tennis you play with your hands and win with your head. According to one celebrated tennis player, the three marks of a high-class tennis player are: the endurance of a long-distance runner, the speed of a sprinter, and the swift thinking of a chess player in time-trouble.

3.2. The arithmetic of tennis. The scoring system in tennis derives from the times when it was played for stakes. In France, the value of a game was a coin of 60 sous; it was changed to four coins of 15 sous, which, evidently, made the value of four strokes: 15, 30, 45 (later abbreviated to 40), and 60.

So, when one of the players wins the first point, the score is 15:0 (or 0:15). Should the same player win the second point, the score becomes 30:0. After winning the third point, his score is 40:0, and after winning the fourth, it is 60:0, and the player takes the game.

Should one of the players, after winning the first point, lose the second, 15 is scored to his opponent, and so on. Hence, the score (the server's score is always called first) in a game can only be one of the following: 15:0, 30:0, 40:0, 0:15, 0:30, 0:40, 15:15, 30:15, 40:15, 15:30, 15:40, 30:30, 40:30, 30:40, deuce, advantage in, advantage out, game.

Deuce is a tie in points, starting with the sixth point; advantage in/out starts with the seventh point, if the server scores/loses the point after deuce. The server takes the game, if, following advantage in, he scores the next point;

[3] A ball game similar to rounders. (Translator's note)

the receiver wins if the server loses the point following advantage out.

After the first game is over, the second game starts, in which service passes to the opposite side, and so on, until a set is completed. This happens when one side wins at least six games, with a margin of at least two games. Hence, a set is completed as soon as the score is one of the following: 6:0, 6:1, 6:2, 6:3, 6:4, 7:5, 8:6, and so on. One set follows another until one side wins the match by taking two sets out of three or three out of five, depending on the terms of the competition. Should one side win two (three) consecutive sets, that side wins, and the remaining sets are not played. Hence, the winning score may be 2:0, 2:1 (3:0, 3:1, 3:2).

Well, dear reader, you have learned enough about the basic rules of tennis to be ready for its further consideration (for more detailed information about the game, see **[14]**). The game is remarkable for the diversity of strokes, which may be varied in direction, force, length and height; the ball may be made to spin on differently oriented axes, and so on. Tennis can be played by opponents of different ages and fitness levels; for example an older player, by virtue of good stroking and placing technique and longer experience, can outplay a more agile and vigorous younger opponent. Therefore, age need not hinder a player from being a competition enthusiast, not to mention the fact that finding a suitable partner for a social game is not a problem.

But to return to the principles of the play and scoring, for all their simplicity, they are thought out very thoroughly.

Indeed, the stipulation that there be a two-point margin, whether in a game or a set, gives both sides an equal chance of success (consider the attacking character of service, which alternates between server and receiver after each game, as well as the fact that the ball is served to alternate areas). Furthermore, each exchange is important, if not decisive, to the outcome of the match, tipping the scale in favor of one side or the other.

3.3. How long should one play? Now, we can start playing: you (READER) and one of the writers (WRITER). Should the WRITER be a much better player than the READER, his superiority will be apparent soon enough—in the very first set. If, however, the difference is not so great, the score in each game and set will be unstable. That exactly is the case in matches between members of the world's tennis elite, for example, on the Grand Prix circuit or in unofficial world championship tournaments at Wimbledon (England), open to both professionals and amateurs.

The world's best players are ranked according to their performance (for a special discussion of ranking see §9).[4] They all have their merits (and hardly any weak points) and meet as equals, while the outcome usually depends on additional circumstances such as mental readiness, physical condition, the emotional pitch, the court (clay, grass, plastic, etc.), the weather, and so on.

What, then, should be the structure of a match between players of approximately equal skill, for one of them to be victorious? Naturally, the longer

[4] Rankings can be national (regional) for players in different age groups. The all-Union (of women, men, and juniors), republic, Moscow, and other rankings are regularly renewed.

the match, the more manifest will be the superiority of one of the opponents. There are no draws in tennis; the game goes on until one of the opposing sides wins. However, even the best of players do not possess unlimited energy. Experience shows that five-set (and even three-set) matches often are three or four hours long.

Here are a few examples. In the 1976 Davis Cup semifinals, the Soviet team leader Alexander Metreveli defeated Manuel Orantes, seeded number two in 1975, in the fifth set of a four-hour match.

The match between Bjorn Borg (a Swedish player who won Wimbledon five times in a row) and John McEnroe, U.S., in the Wimbledon finals in 1979 was 3 hours and 30 minutes long and was won by McEnroe, ranked number one in 1981. Throughout the match a high tempo was maintained—up to 66 stokes a minute (with McEnroe serving). In 1980, the finals were won by Borg. In 1983 and 1984, McEnroe won the championships again. In four years (1978–1981) these outstanding players met 18 times, each scoring 9 victories.

Up to 1983, the matches between the world's two top female players, Chris Evert Lloyd and Martina Navratilova, were won 29 times by Lloyd and 31 times by Navratilova.

In the 1983 Wimbledon semifinals, Curran and Lewis (ranked 15th and 91st, respectively) played 61 games, which took four hours, to decide the outcome of the match. In the deciding fifth set Lewis had luck on his side. In the finals, McEnroe beat him in three sets.

To reach the finals of the French Open Championship in 1981, Borg played altogether 14 hours and 8 minutes, and his opponent, Ivan Lendl, 17 hours and 21 minutes. They played 41 games in the final five-set match.

In addition to a match's indeterminate length, players often have to play singles, doubles, and mixed doubles on the same day. That was the reason for the introduction of the tiebreaker, where a set is not continued until one of the sides achieves a two-game margin. At the score of 6:6, a tiebreaker (the deciding thirteenth game) is played in which scoring is different from a regular game: to win it, one must be the first to get seven points, with a margin of at least two points.[5]

In other words, each successful exchange yields a point. The side that gets seven points wins the thirteenth game and the set, if the opponent has scored no more than five points. Otherwise, the game continues until one side acieves a two-point margin. Although theoretically the contest can go on indefinitely, in practice it ends quickly enough.

3.4. Elementary concepts of probability theory. Let us construct a mathematical model of tennis and make clear, with its aid, some questions concerning the structure of a tennis match.

First of all, what should we adopt as the measure of a player's perfor-

[5] Scoring under the tie-breaking system was proposed by J. Van Allen (U.S.) and first tested in 1970, at the U.S. Open in Philadelphia. It has since been introduced in many countries, including the U.S.S.R.

mance? Obviously, it is the share of the points he wins on the average. Here we shall have to resort to some concepts and basic principles of probability theory. You can read about them in [7, **8**]. Still, we shall recall them.

Consider an individual exchange as a trial J. The trial may have for the player two mutually exclusive outcomes: he either wins the point (event A) or loses the point (even B).

The *frequency of the random event* A (B, respectively) in n trials is the ratio m/n, where m is the number of trials in which event A (B) occurred, to their total number n. Naturally enough, different exchanges proceed under different conditions. But even if the conditions were the same, it would be difficult to derive a regularity from the results of a single exchange (trial J). On the other hand, experiments show that when we consider each sufficiently long sequence of n trials, the frequency m/n of the event A is little different from some value $P(A)$. This fact demonstrates the so-called statistical stability of frequency. The value $P(A)$ is the *probability* of the event A.

The larger the number of trials n, the less the frequency m/n deviates from probability $P(A)$. This repeatedly verified fact has mathematical confirmation in the Bernoulli theorem (one form of the law of large numbers) (see [**3**, p. 69]). That is why, with a large number of trials, the frequency m/n is taken for the approximate value of the probability $P(A)$. Note that always $0 \leq P(A) \leq 1$.

Let us assume that for each player we know the probability $P(A)$ that he wins a point and the probability $P(B)$ that he loses a point.[6]

Naturally,

$$P(A) + P(B) = 1. \tag{1}$$

The *sum* (*union*) $A + B$ (or $A \cup B$) of events is an event that is realized at outcomes resulting in A or resulting in B (or both). Furthermore, the outcomes resulting in A and B simultaneously, are counted only once.

The *product* (*intersection*) AB (or $A \cap B$) of two events is an event realized if and only if we have outcomes resulting both in A and B.

Events A and B are *mutually exclusive* if their product is an impossible event: its probability is 0.

In the present case, a trial J results in two mutually exclusive outcomes (win or loss). Their sum $A + B$ is a certain event—its probability is 1: $P(A + B) = 1$; and the product AB is an impossible event: $P(AB) = 0$.

Formula (1) is a special case of the *probability addition theorem*: if the outcomes A and B of a trial J are mutually exclusive, then the probability of the sum $A + B$ of the outcomes A and B is equal to the sum of probabilities of these outcomes: $P(A + B) = P(A) + P(B)$.

The theorem of addition of probabilities is generalized for the case when a trial results in any finite number $B_1, \dots, B_k$ of mutually exclusive outcomes

---

[6] In this instance probability $P(A)$ is, of course, relatve: it depends on whom the player has for his opponent.

(i.e., each product B_iB_j, given $i \neq j$, is an impossible event):

$$P(B_1 + B_2 + \cdots + B_k) = P(B_1) + P(B_2) + \cdots + (B_k).$$

Important to the further discussion is the concept of the conditional probability $P(A/B)$ of event A, provided that the event B occurs: *conditional probability* $P(A/B)$ is the ratio of the number of outcomes of the trial J having resulted in A, which also result in B, to the number of all outcomes resulting in B. It follows from the definition that $P(A/B) = P(AB)/P(B)$.

The event A is *independent* of the event B if the conditional probability $P(A/B)$ is equal to the unconditional probability $P(A)$, i.e., $P(A/B) = P(A)$. From the above it follows that the *theorem of multiplication of probabilities* is correct for independent events:

$$P(AB) = P(A)P(B).$$

For dependent events

$$P(AB) = P(A)P(B/A) = P(B)P(A/B).$$

Lastly, we recall the *formula of total probability* needed further on. Let events $B_1, \dots, B_k$ be mutually exclusive, and let the event A occur when at least one of the events $B_1, \dots, B_k$ occurs. Then the following identity is correct:

$$A = A(B_1 + \cdots + B_k) = AB_1 + \cdots + AB_k,$$

and the total probability formula is

$$P(A) = P(AB_1) + P(AB_2) + \cdots + P(AB_k)$$

or

$$P(A) = P(B_1)P(A/B_1) + P(B_2)P(A/B_2) + \cdots + P(B_k)P(A/B_k).$$

3.5. The model of a game is a Markov chain. Now let us construct a mathematical model of a game of tennis between the WRITER (W) and the READER (R), assuming that we know the probabilities $P(\text{W})$ and $P(\text{R})$ of the WRITER and READER, respectively, winning the point. To specify, let $P(\text{R}) = 0.4$; $P(\text{W}) = 0.6$ (the WRITER plays a little better than the READER). It is no accident that $P(\text{R}) + P(\text{W}) = 0.4 + 0.6 = 1$ (when lost by one side, the point is won by the other). Figure 2 shows how the score in a game can change. The numbers next to the arrows indicate the probability of the corresponding change in the score. For example, at 15:15, the WRITER may win the point with probability 0.6, so the score will be 30:15; and the READER may win with probability 0.4, so the score will then be 15:30.

Let us say that we have a *system*—a game of tennis. The *states* of this system are defined by the score. Transition from one state (or score) to the next depends only on the present state and, of course, on the transition probabilities (or the numbers next to the arrows) but not on any preceding states.[7] Note that any system for which transition from one state to another depends only on its present state but not on the previous history of the

[7] We are somewhat idealizing the situation, leaving out of account some other circumstances such as the factor of service, psychological factors, or adaptation to the opponent's style, i.e. "learning" in the course of the game.

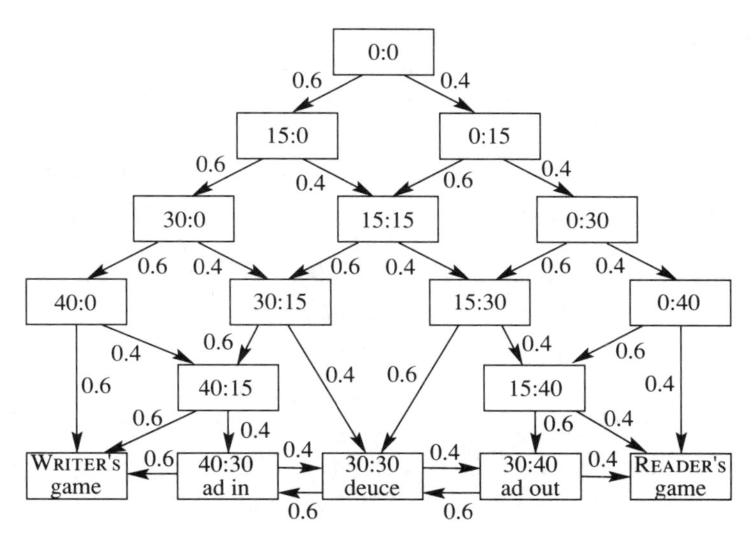

FIGURE 2

process is known in probability theory as the *Markov chain* (A. A. Markov, 1856–1922). In the general case a finite Markov chain can be given as a geometric diagram (or *oriented graph*) in which rectangles (*vertices of the graph*) represent states and arrows (*edges of the graph*) indicate transitions from state to state. Marked next to each arrow is the probability of the respective transition. Figure 2 is an example of a graph of a finite Markov chain, descrbing the states of a system—a game of tennis.

Note that in Figure 2 the scores 40:30, 30:30, and 30:40 are, understandably, joined with "ad in", "deuce", and "ad out", respectively.

The Markov chain allows the existence of states of different types [25]. First, there is the *nonrecurrent state* or one a system cannot resume once it has left it. In the present case such states are fairly numerous, for example, 15:30 or 40:40, and so on. Second, there is the *recurrent state*, or any state other than nonrecurrent. Examples of these states are "ad in", "deuce", and "ad out". Another important state is the *absorbing state.* We call a state absorbing if a system having acquired it cannot change into any other states. "WRITER's Game" and "READER's Game" are examples of an absorbing state.

3.6. Play! So, the score is 0:0. The WRITER serves. The probability of the score being 15:0 is 0.6, and of it being 0:15 is 0.4. Let us find the probability of transition from 0:0 to 30:0, 15:15, and 0:30. The score 30:0 occurs when the WRITER's wins two successive exchanges with probability $0.6.\cdot 06 = 0.36$ (by the multiplication theorem). The probability of 0:30 after two exchanges is $0.4 \cdot 0.4 = 0.16$. The score 15:15 occurs if the WRITER wins the first exchange and the READER wins the second exchange or vce versa. Let H_1 be the hypothesis that the WRITER wins the first exchange, and let H_2 be the hypothesis that the READER wins the first exchange. Then $P(H_1) = 0.6$

is the probability that the first hypothesis is true, and $P(H_2) = 0.4$ is the probability that the second hypothesis is true. Consider the random event Q that the score is 15:15, given the conditional probabilities $P(Q/H_2) = 0.6$ and $P(Q/H_1) = 0.4$. From the total probability formula we find that

$$P(Q) = P(H_1)P(Q/H_1) + P(H_2)P(Q/H_2) = 0.6 \cdot 0.4 + 0.4 \cdot 0.6 = 0.48.$$

Thus, score probabilities after two exchanges are: $P(30:0) = 0.36$, $P(15:15) = 0.48$, and $P(0:30) = 0.16$. We now find the probabilities of the possible scores after three exchanges. It is easy to see that

$$P(40:0) = 0.6 \cdot 0.6 \cdot 0.6 \approx 0.22,$$
$$P(0:40) = 0.4 \cdot 0.4 \cdot 0.4 \approx 0.06.$$

We find the probabilities of the other two possible scores by the total probability formula just as we did when the score was 15:15, viz.,

$$P(30:15) = P(30:0) \cdot 0.4 + P(15:15) \cdot 0.6 \approx 0.43,$$
$$P(15:30) = P(15:15) \cdot 0.4 + P(0:30) \cdot 0.6 \approx 0.29.$$

Let us generalize the results: in order to find the probability of the score indicated in some rectangle in Figure 2, we must find the sum of the products of the probabilities shown next to the arrows entering the rectangle and the probabilities of the scores indicated in the rectangles these arrows exit.

3.7. Finishing the game. After four or five exchanges, one of the states indicated on the bottom line of Figure 2 must occur. Applying the rule we already know, we find that after four exchanges the probabilities that these states will occur are as follows: $P(\text{WRITER's game}) \approx 0.13$; $P(40:15) \approx 0.35$; $P(30:30) \approx 0.35$; $P(15:40) \approx 0.15$; $P(\text{READER's game}) \approx 0.02$. After five exchanges their values will be equal: $p_1^0 = P(\text{WRITER's game''}) = 0.6^4(1 + 4 \cdot 0.4) \approx 0.33$, $p_2^0 = P(\text{``ad in''}) = 4 \cdot 0.6^3 \cdot 0.4^2 \approx 0.15$, $p_3^0 = P(\text{``deuce''}) = 6 \cdot 0.6^2 \cdot 0.4^2 \approx 0.33$; $p_4^0 = P(\text{``ad out''}) = 4 \cdot 0.6^2 \cdot 0.4^3 \approx 0.10$; $p_5^0 = P(\text{``READER's game''}) = 0.4^4(1 + 4 \cdot 06) \approx 0.09$.

Subsequently the situation gets somewhat more complicated due to the possibility of a *random walk* (in three states) [8] or, simply, playing for "ad in-ad out." Therefore in order to make finally clear the probabilities of the game being won by the WRITER and the READER, respectively, let us consider separately the bottom line in Figure 2.

Let us compile the following table.

	1	2	3	4	5
1	1	0	0	0	0
2	0.6	0	0.4	0	0
3	0	0.6	0	0.4	0
4	0	0	0.6	0	0.4
5	0	0	0	0	1

[8] A highly interesting problem of a symmetric random walk is to be found in [9].

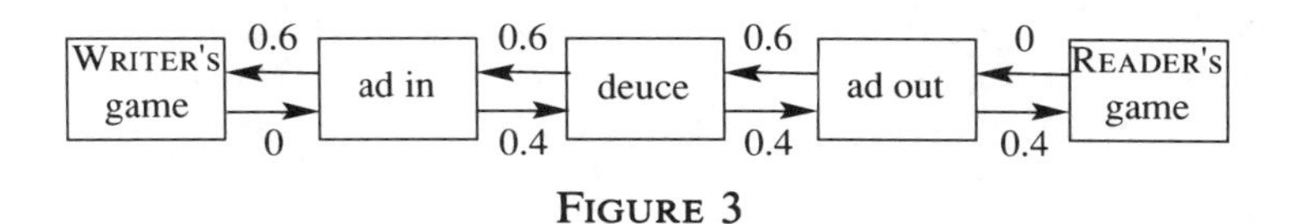

FIGURE 3

Indicated at the intersection of the ith row and the jth column in the table is the probability of transition from a state j to a state j. For example, 1 at the intersection of the first row and first column means that the "WRITER's game" is an absorbing state, i.e., the game is finished and its score cannot be changed. The intersection of the third row and second column is marked 0.6, i.e., there is a 0.6 probability that the score will change from "deuce" to "ad in", 0.4 at the intersection of the same row and the fourth column indicates the probability that the score will change from "deuce" to "ad out". Naturally, the sum of the probabilities noted on the same row equals 1, since each time a point is played by the score must change in favor of one side.

Let us represent our table as a matrix:

$$T = \begin{pmatrix} 1 & 0 & 0 & 0 & 0 \\ 0.6 & 0 & 0.4 & 0 & 0 \\ 0 & 0.6 & 0 & 0.4 & 0 \\ 0 & 0 & 0.6 & 0 & 0.4 \\ 0 & 0 & 0 & 0 & 1 \end{pmatrix}.$$

The matrix T is the *transition matrix* of the Markov chain shown in Figure 3. We shall take the state probabilities after five exchanges as the components of the vectors $\mathbf{p}^0 = (p_1^0, p_2^0, p_3^0, p_4^0, p_5^0)$, and we shall call it the *initial* (pre-random walk) *probability distribution vector* of the respective states. In our game the numerical values of p_i^0 $(i = 1, \dots, 5)$ have already been computed.

3.8. Vector operations. Let $\mathbf{x} = (x_1, \dots, x_5)$ be a five-dimensional vector, and let A be a given matrix of order five. The product of the vector $\mathbf{x}$ by the matrix A is (by definition) the row vector $\mathbf{x}' = (x_1', \dots, x_5')$ in which each coordinate x_i' is equal to the scalar product of the vector $\mathbf{x}$ by the ith column vector of the matrix A. For example, multiplying the vector $\mathbf{x}$ by our transition matrix T we shall have

$$\begin{aligned} x_1' &= x_1 \cdot 1 + x_2 \cdot 0.6 + x_3 \cdot 0 + x_4 \cdot 0 + x_5 \cdot 0, \\ x_2' &= x_1 \cdot 0 + x_2 \cdot 0 + x_3 \cdot 0.6 + x_4 \cdot 0 + x_5 \cdot 0, \\ x_3' &= x_1 \cdot 0 + x_2 \cdot 0.4 + x_3 \cdot 0 + x_4 \cdot 0.6 + x_5 \cdot 0, \\ x_4' &= x_1 \cdot 0 + x_2 \cdot 0 + x_3 \cdot 0.4 + x_4 \cdot 0 + x_5 \cdot 0, \\ x_5' &= x_1 \cdot 0 + x_2 \cdot 0 + x_3 \cdot 0 + x_4 \cdot 0.4 + x_5 \cdot 1. \end{aligned}$$

Suppose, further that $\mathbf{p}^0 = (p_1^0, \dots, p_5^0)$ is our initial probability distribution for the states represented in Figure 3 and T is the matrix of transition probabilities. What is the probability that after the first step (the very next exchange) the score will be, say "deuce"?

The state "WRITER's game" will change into "deuce" with probability 0 (the game is finished); "ad in" will change with probability 0.4; "deuce" will change with probability 0; "ad out" will change with probability 0.6; and "READER's game" will change into "deuce" with probability 0. By the total probability formula we find that after the first step

$$p_3^1 = P(\text{"deuce"}) = p_1^0 \cdot 0 + p_2^0 \cdot 0.4 + p_3^0 \cdot 0 + p_4^0 \cdot 0.6 + p_5^0 \cdot 0.$$

Thus $p_3^{(1)}$ turns out to be the scalar product of the initial distribution vector $\mathbf{p}^0$ by the third column of the matrix T.

Using similar reasoning with respect to the other four states we conclude that after the first exchange the probabilities that new states will occur can be found as respective components of the vector

$$\mathbf{p}^0 T = \mathbf{p}^{(1)} = (p_1^{(1)}, p_2^{(1)}, p_3^{(1)}, p_4^{(1)}, p_5^{(1)}).$$

Repeating the same operations with the vectors $\mathbf{p}^0 T = \mathbf{p}^{(1)}$, $\mathbf{p}^{(2)} = \mathbf{p}^{(1)} T, \ldots$ we obtain the result that after n exchanges the respective probabilities will be the components $p_1^{(n)}, p_2^{(n)}, p_3^{(n)}, p_4^{(n)}, p_5^{(n)}$ of the vector[9]

$$\mathbf{p}^{(n)} = \mathbf{p}^{(n-1)} T = (\mathbf{p}^0 T) T \cdot \dots \cdot T = \mathbf{p}^0 T^n.$$

It is possible to find[10] what are known as limiting probabilities p_1^* and p_5^* (the probabilities of the game being won by the WRITER and the READER for unlimited growth of n), which in our example are equal to 0.736 and 0.264, respectively.

Note that for arbitrary probabilities p and q $(p + q = 1)$ of an exchange being won by the WRITER and the READER, respectively, it is possible, by the same argument, to establish that after four or five exchanges the probabilities preceding a random walk are:

$$p_1^0 = p^4(1 + 4q); \; p_2^0 = 4p^3 q^2; \; p_3^0 = 6p^2 q^2; \; p_4^0 = 4p^2 q^3; \; p_5^0 = q^4(1 + 4p).$$

For example, if $p = q = \frac{1}{2}$, then $p_1^0 = \frac{3}{16}$; $p_2^0 = \frac{1}{8}$; $p_3^0 = \frac{3}{8}$; $p_4^0 = \frac{1}{8}$; $p_5^0 = \frac{3}{16}$, and the vector p^0 is $\mathbf{p}^0 = (\frac{3}{16}, \frac{1}{8}, \frac{3}{8}, \frac{1}{8}, \frac{3}{16})$.

3.9. Now let us go on to complete the set. So, we have found the probabilities of each side winning a game. We shall now compute the probabilities of their winning a set. To this end, we shall write out all the possible changes in the score within a set and represent them as an oriented graph (Figure 4).

Looking at the new finite Markov chain (Figure 4), we see that after 11 or 12 games there occurs a random walk. It is due to the fact that to win a set one must have a margin of at least two games. The graph of this random walk is shown in Figure 5.

---

[9] T^n is the matrix T raised to nth power.

[10] In the theory of Markov chains (given some assumptions on the transition matrix T) there is a proof of the existence of limiting probabilities.

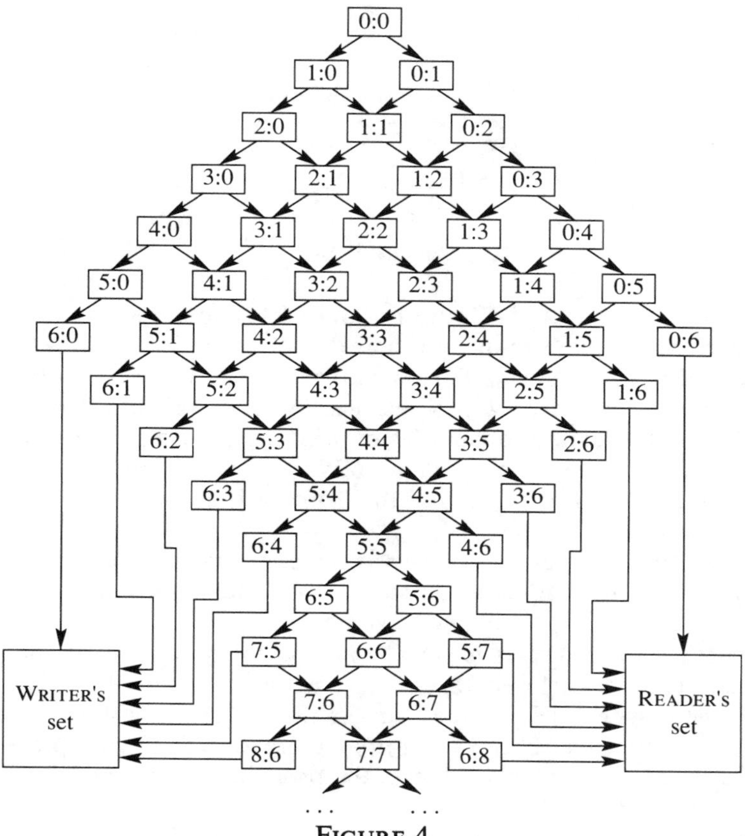

FIGURE 4

FIGURE 5

The procedure by which we computed the probabilities of winning a game can be used to find the probabilities of winning a set:

$$P(\text{WRITER's set}) = 0.966, \qquad P(\text{READER's set}) = 0.034.$$

As we see, the probability that the WRITER wins the set is close to 1. This is only to be expected. After all, the probability of the WRITER winning a point is one and half times that of the READER. The probability that the WRITER wins a three-set match is 0.996 (the reader can verify it), and the probability that he wins a five-set match is 0.9996 or practically a certainty. It is therefore clear that it would be of no purpose in this case to play more than three sets.

Now suppose that the players are of about the same strength and the probability that the WRITER wins a point is 0.51, and that the READER wins a

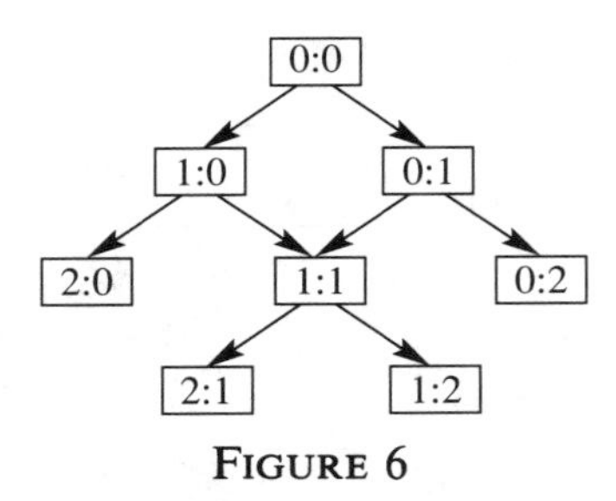

FIGURE 6

point is 0.49. In other words, out of a hundred points the READER wins on the average merely two points less than the WRITER. In this case the probability that the WRITER wins the set is 0.573, and the probability that the READER wins the set is 0.427. Thus, while the probabilities of winning a point differ by 0.02, the probabilities of winning a set differ by seven times as much. Nevertheless in one set the odds in favor of the stronger player are not very great; out of ten games the WRITER is going to win six, and the READER four, on the average.

The probability that each player wins one set, i.e., that the score is 1:1, is as high as 0.488 (the reader can calculate it from the graph in Figure 6).

3.10. Let's have it out! It is clear from the foregoing that it will take at least three sets for two closely ranked opponents "to have it out" in a match. But even then (with $P(\mathrm{W}) = 0.51$ and $P(\mathrm{R}) = 0.49$) the picture still will not be very convincing. The WRITER may score 2:1 or 2:0 with probabilities 0.255 and 0.328, respectively. Hence, the probability that the WRITER wins the match is 0.583.

The probability of winning a five-set match is already 0.625 for the WRITER, and 0.375 for the READER. And though the difference in this instance is 12.5 times that between the probabilities of winning a point, it might still be desirable to raise the number of sets to seven. This, however, would make matches too strenuous and far too long.[11] An average set consists of nine or ten games. A game is usually three or four minutes, sometimes up to ten minutes long. A three-set match is on the average 2 hours and 30 minutes long, and a five-set match is four hours long.[12]

Now we can see that besides tradition there are also certain theoretical reasons why major final, semifinal, and sometimes other matches consist of five sets (less often, of three sets).

There also are theoretical grounds for scattering (seeding) the top players over the draw sheet to avoid matching them in the early rounds (see 8.1, The Olympic System). Early on in a tournament top seeds and their lower-ranked opponents meet in three-set matches. As things move toward the finish, the

[11] Women and juniors play only three-set matches.

[12] Besides, with players of about equal strength, even a seven-set match would not significantly change the probablity of the outcome; the number of matches in which the winner is decided is the fifth set amounts to 20–30%.

tension mounts, the players are more evenly matched, and to determine the winner they change over to five-set matches.

Until 1976 on the Grand Prix circuit the number of three- and five-set match tournaments was about even (34 and 38, respectively). By 1982, three-set matches were played in 62 out of 93 tournaments.

3.11. Practice makes perfect. In the course of a game every tennis player takes stock of his technical and tactical errors and adapts to his opponent's style; he learns and improves his performance.

This can be reflected by a model also constructed as a Markov chain. It can be assumed that in the Markov chains describing the contesting of a game (or a set), the probabilities of transition from one state to the next change in keeping with certain laws.

Let us begin with a tiebreaker.

Suppose that a player (WRITER), after winning a point (state S_1), wins the next point with probability $p = 0.5$ (he is as alert and nimble as he normally is). But after losing a point (state S_2) the WRITER is more determined, more alert, and he tries harder. Therefore, the probability that he wins the next point may be higher, say $q = 0.7$. Thus we have the following transition matrix from S_i to S_j $(i, j = 1, 2)$:

$$T = \begin{pmatrix} 0.5 & 0.5 \\ 0.7 & 0.3 \end{pmatrix} \begin{matrix} S_1 \\ S_2 \end{matrix} \qquad \text{columns: } S_1 \quad S_2$$

and the corresponding state graph (Figure 7).

At six games each the player whose turn it is to serve serves the first point of the tiebreaker. His opponent serves the second and third points. Then each player serves twice in turn until the game (and so the set) is completed. Assume that the WRITER starts the tiebreaker by winning the first point. The probability that he wins the second point in our example is $p_1 = 0.5$, and the probability that he loses is $q_1 = 0.5$. We shall consider that (p_1, q_1) is the initial probability distribution. What is the probability that he wins the third point and that he loses it?

Let us denote the respective probabilities by p_2 and q_2. We shall find $p_2 = 0.5^2 + 0.5 \cdot 0.7 = 0.60$ and $q_2 = 0.5^2 + 0.5 \cdot 0.3 = 0.40$. In other words, the distribution (p_2, q_2) of probabilities of the outcome of the third game is given by the product of the initial distribution (p_1, q_1) and the matrix T:

$$(p_2, q_2) = (p_1, q_1) \begin{pmatrix} 0.5 & 0.5 \\ 0.7 & 0.3 \end{pmatrix}.$$

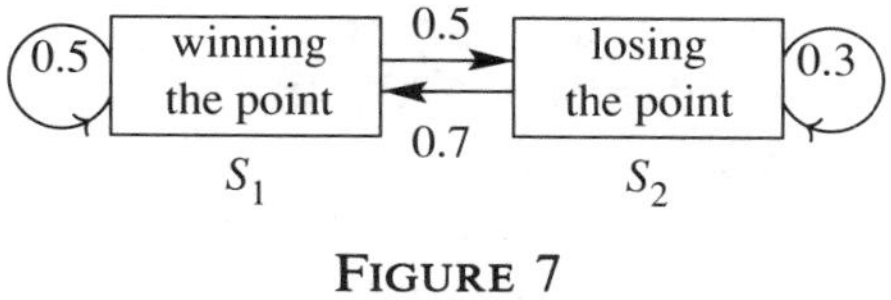

FIGURE 7

By the same reasoning we shall find that the probability distribution of the outcome of the $(n+1)$th point is

$$(p_n, q_n) = (p_{n-1}, q_{n-1})T = (p_1, q_1)T^{n-1}.$$

Specfically, $p_3 = p_2 \cdot 0.5 + q_2 \cdot 0.7 = 0.580$; $q_3 = p_2 \cdot 0.5 + q_2 \cdot 0.3 = 0.420$; $p_4 = 0.584$; $p_5 = 0.5832$, $p_6 = 0.58336$; $q_4 = 0.416$; $q_5 = 0.4168$; $q_6 = 0.41663$; and so on. Lastly, we can identify the limiting probability values:

$$p^* = \lim_{n\to\infty} p_n = \tfrac{7}{12},$$
$$q^* = \lim_{n\to\infty} q_n = \tfrac{5}{12}.$$

Let us now suppose that the WRITER started the tiebreaker by losing the point. In this case, the initial distribution will be $p_1' = 0.7$, $q_1' = 0.3$, and the distribution of outcomes in playing the third point will be

$$p_2' = 0.7 \cdot 0.5 + 0.3 \cdot 0.7 = 0.56,$$
$$q_2' = 0.7 \cdot 0.5 + 0.3 \cdot 0.3 = 0.44.$$

For the $(n+1)$th point we obtain the distribution

$$(p_n', q_n') = (p_{n-1}', q_{n-1}')T = (p_1', q_1')T^{n-1}.$$

Specifically,

$$p_3' = 0.588; \qquad p_4' = 0.5824; \qquad p_5' = 0.58352;$$
$$q_3' = 0.412; \qquad q_4' = 0.4176; \qquad q_5' = 0.41648;$$

$$p_6' = 0.583296; \qquad q_6' = 0.416706.$$

The limiting probabilities resume their prevous values

$$p'^* = \lim_{n\to\infty} p_n' = \tfrac{7}{12}, \qquad q'^* = \lim_{n\to\infty} q_n' = \tfrac{5}{12}.$$

3.12. Model of a tiebreaker. Thus, regardless of which side wins the first point in a tiebreaker, the probabilities of which side will win the further points soon settles down, so that after the third or fourth point they hardly differ. That is why from the moment the score reaches six–all and further in the game, until one of the opponents achieves the two-point margin, the probabilities of either the WRITER or the READER winning the next point may be considered constant and equal to $\frac{7}{12}$ and $\frac{5}{12}$, respectively.

The graph of changes in the point score in a tiebreaker is identical to that of changes in the game score in a set (Figure 4). The only difference is in the numerical values of transition probabilities.

Utilizing the graph shown in Figure 4 we can calculate (similar to a game—see §3.7) the probability distribution $\mathbf{p}^0 = (p_1^0, p_2^0, \dots, p_5^0)$ that after 11 or 12 points each of the following five states is achieved: WRITER's tiebreaker, WRITER's ad in, deuce, WRITER's ad out (or READER's ad in), READER's

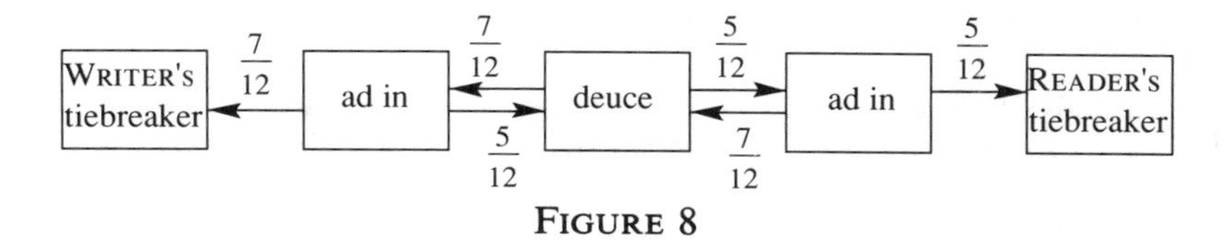

FIGURE 8

tiebreaker. A graph of the Markov chain of these five states is shown in Figure 8.

We shall take $\mathbf{p}^0$ as the initial distribution for a random walk. Let us introduce the transition matrix of the chain under consideration:

$$T = \begin{pmatrix} 1 & 0 & 0 & 0 & 0 \\ \frac{7}{12} & 0 & \frac{5}{12} & 0 & 0 \\ 0 & \frac{7}{12} & 0 & \frac{5}{12} & 0 \\ 0 & 0 & \frac{7}{12} & 0 & \frac{5}{12} \\ 0 & 0 & 0 & 0 & 1 \end{pmatrix}.$$

Then the probability distribution $\mathbf{p}^1$ that the above five states are achieved after playing the next point will be found as the product of the vector $\mathbf{p}^0$ by the matrix T: $\mathbf{p}^1 = \mathbf{p}^0 T$. The probability distribution after another point will be $\mathbf{p}^2 = \mathbf{p}^1 T$, and so on, untl the values of the coordinates of the probability distribution vector $\mathbf{p}^n = \mathbf{p}^{(n-1)} T$ become settled.

To expand this situation a little, suppose that the WRITER's opponent, the READER, also lays himself out after he has lost the point, and assume that the probability of his winning the next point is, say 0.6, rather than 0.5. In that case the transition matrix will be as follows:

$$T' = \begin{pmatrix} 0.4 & 0.6 \\ 0.7 & 0.3 \end{pmatrix}.$$

Thus it is possible to allow to some extent for psychological, emotional, and similar other factors.

The reader can compute for this situation the limiting probabilities p^* and q^* with which both players will arrive at the final part of the tiebreaker.

3.13. Markov chains and basketball. Let us try to generalize and formalize the apparatus we used when we considered the intricacies of a tennis match. Such a mathematical description of situations that evolve in the form of a random process is based on the motion of a Markov chain, in which the next state depends on the present but not on the past one.

A random process is called discrete of the transition of the system from one state to the next is only possible at strictly predetermined fixed moments of time $t_1, t_2, \dots, t_n, \dots$. In the time intervals between two consecutive moments the state of the system remains unchanged.

A random process is called continuous if the transition of the system from one state to another is possible at any undetermined random moment of time t.

We first consider Markov random processes with discrete states and discrete time.

Let there be given a physical system S with possible states $S_1, S_2, \dots, S_n$, while the transitions ("jumps") from one state to another are possible only at moments $t_1, t_2, \dots, t_k, \dots$.

We shall call these moments "steps" or "stages" of the process and consider the random process taking place in the system as a function of the integer-valued argument $1, 2, \dots, k, \dots$ (the index of the step).

A random process taking place in the system consists of having the system S in one state or another at the consecutive moments of time $t_1, t_2, \dots$; for example, the system S may behave live this:

$$S_1 \to S_3 \to S_5 \to S_{14} \cdots,$$

or like this:

$$S_1 \to S_2 \to S_1 \to S_3 \to \cdots.$$

In general, the system may not only change states from one moment of time to the next, but also remain in the previous state.

Let us denote by $S_i^{(k)}$ the event: after k steps the system is in state S_i. For any k the events

$$S_1^{(k)}, S_2^{(k)}, \dots, S_i^{(k)}, \dots, S_n^{(k)}$$

are mutually exclusive and make the totality of possible events (sample space).

A process taking place in the system may be represented as a sequence (chain) of events; for example,

$$S_1^{(0)}, S_2^{(1)}, S_1^{(2)}, S_2^{(3)}, S_3^{(4)}, \dots.$$

Such a random sequence of events is called a Markov chain if at each step the probability of transition from any state S_i to any state S_j is independent of how and when the system has arrived at the state S_i.

We shall be describing a Markov chain with the aid of the so-called state probabilities. Assume that at some moment of time (after some step) the system S may be in one of the states

$$S_1, S_2, \dots, S_n;$$

i.e., one event from the sample space of mutually exclusive events has taken place:

$$S_1^{(k)}, S_2^{(k)}, \dots, S_n^{(k)}.$$

Let us denote the probabilities of these events as follows:

$$p_1(1) = p(S_1^{(1)});\ p_2(1) = p(S_2^{(1)});$$
$$\dots;\ p_n(1) = p(S_n^{(1)})$$

—the probabilities after the first step;

$$p_1(2) = p(S_1^{(2)});\ p_2(2) = p(S_2^{(2)});$$
$$\dots,\ p_n(2) = (S_n^{(2)})$$

—the probabilities after the second step; and, in general, after the kth step

$$p_1(k) = p(S_1^{(k)});\ p_2(k) = p(S_2^{(k)});\ \dots,\ p_n(k) = p(S_n^{(k)}).$$

It is easy to see that for each k

$$p_1(k) + p_2(k) + \cdots + p_n(k) = 1,$$

since these are the probabilities of mutually exclusive events forming the sample space.

We shall use the term *state probabilities* for $p_1(k)$, $p_2(k), \dots, p_n(k)$; problem: find the state probabilities of the system for any k.

At each step (moment of time $t_1, t_2, \dots, t_k, \dots$ or index $1, 2, \dots, k, \dots$) there are certain probabilities of the system going from any state to any other state (some of them equal to zero of a direct one-step transition is impossible), and also the probability of the system remaining in the same state.

These probabilities are called transition probabilities of the Markov chain.

A Markov chain is called homogeneous if the transition probabilities are independent of the index of the step. Otherwise, the Markov chain is non-homogeneous.

Consider a homogeneous Markov chain. Let the system S have n possible states $S_1, \dots, S_n$. Assume that for each state we know the probability of a one-step transition to any other state (including the probability of the system remaining in the given state). Denote by p_{ij} the probability of one-step transition from state S_i to state S_j; p_{ii} will be the probability of the system remaining in the state S_i. Let us write the transition probabilities p_{ij} in a rectangular table (transition matrix):

$$P = \|p_{ij}\| = \left\| \begin{matrix} p_{11} & p_{12} & \cdots & p_{1j} & \cdots & p_{1n} \\ p_{21} & p_{22} & \cdots & p_{2j} & \cdots & p_{2n} \\ \cdots & \cdots & \cdots & \cdots & \cdots & \cdots \\ p_{il} & p_{i2} & \cdots & p_{ij} & \cdots & p_{in} \\ \cdots & \cdots & \cdots & \cdots & \cdots & \cdots \\ p_{n1} & p_{n2} & \cdots & p_{nj} & \cdots & p_{nn} \end{matrix} \right\|.$$

Some of the transition probabilities p_{ij} may be equal to zero: that would mean that one-step transition from the ith state to the jth is impossible. Along the main diagonal of the transition matrix are the probabilities of the system remaining in the state S_i (not exiting from it).

Using the events $S_1^{(k)}, S_2^{(k)}, \dots, S_n^{(k)}$ introduced above, the transition probabilities p_{ij} can be written as conditional probabilities:

$$p_{ij} = p(S_j^{(k)}/S_i^{(k)}).$$

It follows from this that the sum of the terms in every row of the matrix P must be equal to one, since, whatever the state the system is in before the

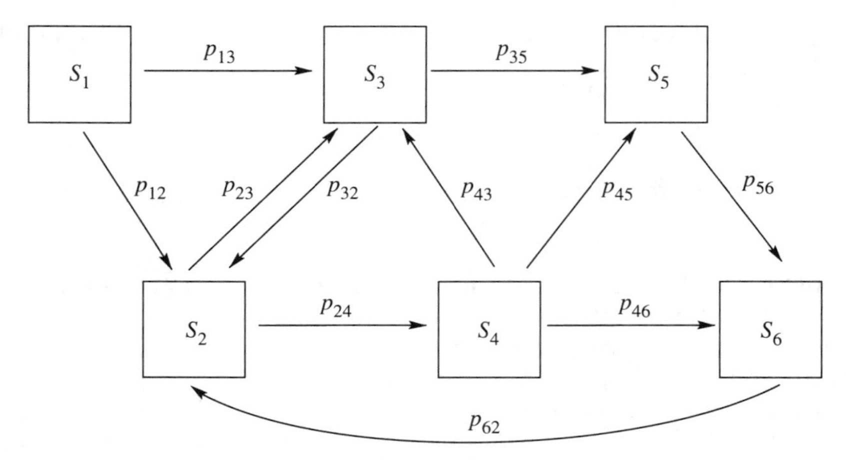

FIGURE 9

kth step, the events $S_1^{(k)}, S_2^{(k)}, \ldots, S_n^{(k)}$ are mutually exclusive and exhaust the sample space.

When considering Markov chains it is often convenient to use graphs representing the states with the corresponding transition probabilities at the arrows, as shown in Figure 9. We shall call such a graph a marked state graph.

Observe that not all transition probabilities are present, only those that are not zero and change the state of the system, i.e., p_{ij} for $i \neq j$; to put on the graph the "delay probabilities" $p_{11}, p_{22}, \ldots$ is unnecessary, since they complement to one the sum of the transition probabilities corresponding to the arrows existing from the given state. For example, for the graph on Figure 9

$$\begin{aligned}
p_{11} &= 1 - (p_{12} + p_{13}), \\
p_{22} &= 1 - (p_{23} + p_{24}), \\
p_{33} &= 1 - (p_{32} + p_{35}), \\
p_{44} &= 1 - (p_{43} + p_{45} + p_{46}), \\
p_{55} &= 1 - p_{56}, \\
p_{66} &= 1 - p_{62}.
\end{aligned}$$

If for some state S_i there are no outgoing arrows (the transition from it to any other state is impossible), the corresponding delay transition p_{ii} is equal to one.

Having at our disposal a marked state graph (or, equivalently, a transition matrix) and knowing the initial state of the system, it is possible to find the state probabilities $p_1(k), p_2(k), \ldots, p_n(k)$ after any (kth) step.

Let us see how this is done.

Assume that at the initial moment of time (before the first step) the system is in some given state, say S_m. Then for the initial moment (0) we have $p_1(0) = 0$, $p_2(0) = 0, \ldots, p_m(0) = 1, \ldots, p_n(0) = 0$; i.e., all state probabilities are equal to zero, except for $p_m(0)$, which is one.

Let us find the state probabilities after the first step. We do know that before the first step the system is in state S_m. So, after the first step it goes into state $S_1, S_2, \ldots, S_m, \ldots, S_n$ with the probabilities, $p_{m_1}, p_{m_2}, \ldots, p_{mm}$, p_{mn}, listed in the mth row of the transition matrix. Thus, the state probabilities after the first step are

$$p_1(1) = p_{m1}; \; p_2(1) = p_{m2}; \ldots ;$$
$$p_i(1) = p_{mm}; \ldots ; p_n(1) = p_{mn}.$$

Let us find the state probabilities after the second step: $p_1(2), p_2(2), \ldots, p_m(2), \ldots, p_n(2)$. We shall calculate them using the total probability formula, assuming that

—after the first step the system is in state S_1;
—after the first step the system is in state S_2;
—after the first step the system is in state S_i;
—after the first step the system is in state S_n.

The probabilities of the hypotheses are known; the conditional probabilities of transition to state S_i for each hypotheses are also known and are written in the transition matrix. By the total probability formula we get

$$p_1(2) = p_1(1)p_{11} + p_2(1)p_{21} + \cdots + p_n(1)p_{n1};$$
$$p_2(2) = p_1(1)p_{12} + p_2(1)p_{22} + \cdots + p_n(1)p_{n2};$$
$$p_i(2) = p_1(1)p_{1i} + p_2(1)p_{2i} + \cdots + p_n(1)p_{ni};$$
$$p_n(2) = p_1(1)p_{1n} + p_2(1)p_{2n} + \cdots + p_n(1)p_{nn};$$

or, in brief,

$$p_i(2) = \sum_{j=1}^{n} p_j(1)p_{ji}, \qquad i = 1, \ldots, n.$$

In this formula the summation is formally extended to all states $S_1, \ldots, S_n$; in fact, one should only account for those for which the transition probabilities p_{ji} are nonzero—in other words, those states from which a transition to S_i (or a delay (rest) in it) is possible.

Thus that state probabilities after the second step are known. Clearly, after the third step they are determined similarly:

$$p_i(3) = \sum_{j=1}^{n} p_j(2)p_{ji},$$

and, in general, after the kth step

$$p_i(k) = \sum_{j=1}^{n} p_j(k_1)p_{ji}.$$

Thus the state probabilities $p_i(k)$ after the kth step are determined from those after the $(k-1)$st step via a recurrence formula; the latter, in their turn, are determined from the state probabilities after the $(k-2)$nd step and so on.

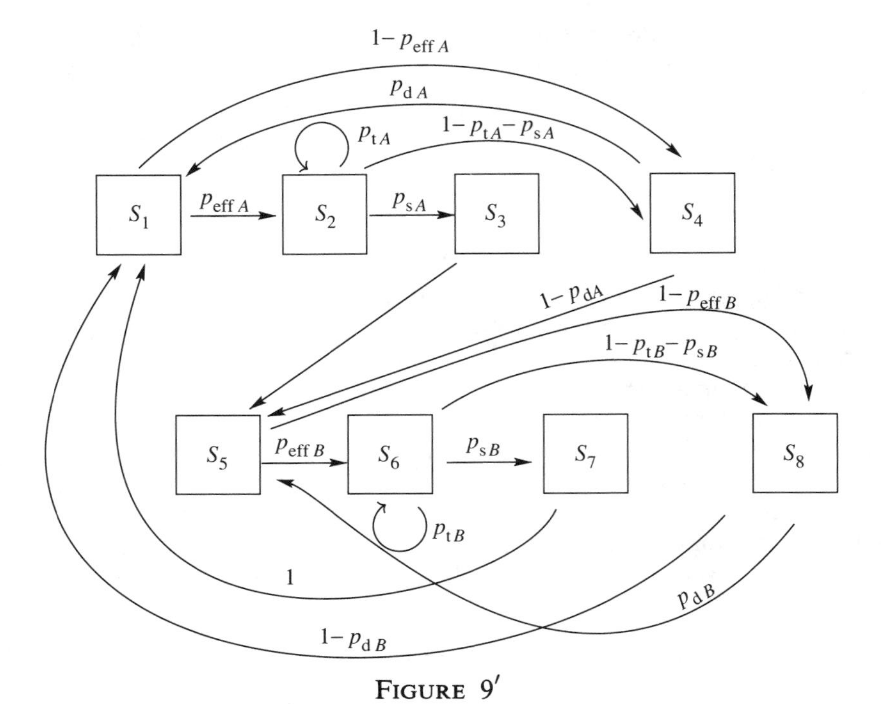

FIGURE $9'$

In matrix form this is written as follows: $\overline{p}(k) = P^k\overline{k}(0)$, where $\overline{p}(k) = (p_1(k), \dots, p_n(k))$, and P is the transition matrix.

If, as $k \to \infty$, there exists $\lim_{k\to\infty} P^k\overline{p}(k) = \overline{p}^*$ independent of the initial probability vector, then $\overline{p}^*$ is called the vector of final probabilities, i.e., these are the probabilities of the system being in one or another state in the steady state mode.

We have already noted that sports competitions offer rich possibilities of applying the apparatus developed.

Consider the following Markov model of a basketball game between two teams A and B; let the state graph of the game be that of Figure $9'$.

Let us comment on the states of the system and the transition probabilities:

S_1—team A goes on attack;
S_2—penetrating attack by team A;
S_3—basket scored by team A;
S_4—ball in possession of team B in defense;
S_5—team B goes on attack;
S_6—penetrating attack by team B;
S_7—basket scored by team B;
S_8—ball in possession of team A in defense;
$P_{\text{eff}A}$, $P_{\text{eff}B}$—the probability of organizing a penetrating attack by team A and team B, respectively;
$P_{\text{d}A}$, $P_{\text{d}B}$—the probability of successful defense of corresponding teams;
$P_{\text{t}A}$, $P_{\text{t}B}$—probability of tip-in;

P_{sA}, P_{sB}—the probability of scoring a basket as a result of a penetrating attack.

Let us try to find the limit probabilities. To this end we consider the relation $\overline{p}(k) = P^k\overline{p}(0)$ and pass to the limit

$$\lim_{k\to\infty} \overline{p}(k) = \lim_{k\to\infty} P^k\overline{p}(0) = \lim_{k\to\infty} P\overline{p}(k-1).$$

Taking into account that $P^k\overline{p}(0) = P\overline{p}(k-1)$, we get that $\overline{p}^* = P\overline{p}^*$.

Thus, to determine the limit probabilities one has to solve the algebraic system of equations

$$\overline{p}^*(P-E) = 0, \qquad \sum_{i=1}^{n} p_i^* = 1,$$

where

$$E = \begin{Vmatrix} 1 & & 0 \\ & \cdots & \\ 0 & & 1 \end{Vmatrix}$$

is the unit matrix.

In our concrete case concerning basketball, we obtain the following system:

$$\begin{aligned}
p_{dA}p_4^* + p_7^* + (1-p_{dB})p_8^* &= p_1^*, \\
p_{\text{eff}\,A}p_1^* + p_{tA}p_2^* &= p_2^*, \\
p_{sA}p_2^* &= p_3^*, \\
(1-p_{\text{eff}\,A})p_1^* + (1-p_{tA}-p_{sA})p_2^* &= p_4^*, \\
p_{sA}p_2^* + (1+p_{dA})p_4^* + p_{dB}p_5^* &= p_5^*, \\
p_{\text{eff}\,B}p_5^* + p_{tB}p_6^* &= p_6^*, \\
p_{sB}p_6^* &= p_7^*, \\
p_1^* + p_2^* + \cdots + p_8^* &= 1.
\end{aligned}$$

By solving this system, we can find the limit probabilities of p_3^* and p_7^*. If we take into account that the length of a basketball game is 40 minutes (amateur rules) and that average time of an attack is 12–15 seconds, it is not hard to understand that there are 80–100 cycles of the process during a game. Consequently, the prognosticated score of the game depends on the final probabilities.

We have conducted a computer modeling of the game; we assumed that $p_{dA} = 1 - p_{\text{eff}\,B}$ and, naturally, $p_{dB} = 1 - p_{\text{eff}\,A}$. The results of the modeling are given in the table on the following page.

We notice that only the parameters of team A have changed.

From the results of the modeling one can draw the following curious conclusions. First, the probability of scoring a basket p_{sA} has a greater influence on the final score of the game than the probability $p_{\text{eff}\,A}$ of effectively organizing an attack by team A. Second, the influence of $p_{\text{eff}\,A}$ in the final score, in turn, is somewhat greater than that of the tip-in (offensive rebound). It

Results of the modeling of the basketball game

$P_{\text{eff}\,A}$	$P_{\text{t}A}$	$P_{\text{s}A}$	$P_{\text{d}A}$	$P_{\text{eff}\,B}$	$P_{\text{t}B}$	$P_{\text{s}B}$	$P_{\text{d}B}$	P_3^*	P_2^*	Prognosticated Score
0.6	0.3	0.7	0.25	0.75	0.3	0.6	0.4	0.099	0.098	99:98
0.6	0.3	0.65	0.25	0.75	0.3	0.6	0.4	0.09	0.102	90:102
0.8	0.3	0.55	0.25	0.75	0.3	0.6	0.2	0.107	0.084	107:84
0.75	0.1	0.6	0.25	0.75	0.6	0.6	0.25	0.074	0.093	74:93
0.7	0.3	0.55	0.25	0.75	0.3	0.6	0.3	0.091	0.096	91:96
0.8	0.2	0.5	0.25	0.75	0.3	0.6	0.2	0.089	0.082	89:82
0.75	0.2	0.5	0.25	0.75	0.3	0.6	0.25	0.081	0.087	81:87
0.75	0.05	0.7	0.25	0.75	0.3	0.6	0.25	0.109	0.074	109:74

is necessary to note, of course, that the results obtained are preliminary and that a more detailed analysis, varying the parameters of both teams, is in order.

4

Those Judges!

The Olympic motto "Faster! Higher! Stronger!" ought to be supplemented today by "More Beautiful!" We know how to measure speed, height, and weight, but how do we measure beauty?

In gymnastics, figure skating and diving (the sports we are about to discuss) judging is done by panels of experts. The judges do not give points only for the difficulty of the exercise, but also for the accuracy, harmony, and beauty of its performance. However, while there are certain fixed standards to which an exercise must conform, as far as its beauty is concerned, every judge has to be guided by his own taste. And, as everybody knows, tastes differ.

People do not often ask why judging in a sport is done one way rather than another. Mostly, it is a matter of tradition. What they do want to know is how judges arrive at their decisions.

Problems that, like judging in sports, are concerned with obtaining experts' opinions are dealt with by a new branch of applied mathematics that studies models and methods of examination by experts, processing of the information provided by experts, and so on. Below we discuss such examinations and their relevance to sports today.

4.1. What is an examination by experts? It is easier to say which of two objects is heavier than to tell how much each of them weighs.

Numerous experiments have shown that people are able to answer qualitative questions—e.g., those involving comparison—more accurately and readily than those requiring quantitative appraisal. Granted, answers to qualitative questions are often based on notions about relations between numbers, i.e., on considerations of a quantitative nature.

An *examination by experts* is a procedure where one group, termed the "decision maker", finds out the opinions of another group, the *experts*, in order to arrive at a decision. Very often, these groups coincide, i.e., the members of a group state their opinions, and the decision is made based on these personal opinions.

One vivid example of examination by experts is the judging of figure skating competitions, where the individual opinions of nine judges are processed to obtain the final result.

The group whose opinion is sought is usually formed of experts in what is to be decided on. Since each expert's judgement is naturally affected by his personal view of the situation, the examination should be arranged so as to minimize the effect of subjective factors.

What kinds of qualitative information can experts be expected to provide?

There are several ways to collect information from a group of experts.

First, the experts may be asked to assess an object on a numerical scale. For example, in gymnastics judges use a ten-point scale, with the points broken down into decimals.

Second, the experts may be asked to assign a place to each object: first place, second, etc. This is called *ranking*.

Third, based on certain characteristics, the experts may break down the total population of the objects into separate classes (subsets). In this case we are talking about classification. As an example we may consider the division of athletes or teams before competitions into groups on the territorial principle or by the sport.

Fourth, it is also possible to compare the objects pairwise, the experts stating which of the two objects they think should be given preference.

These four kinds of qualitative information are basic to examinations by experts. However, other kinds of information can also be obtained from the experts when dealing with specialized problems.

An expert's judgement, whether an evaluation expressed in points or a ranking, is usually called a *relation*, as a more general term.

The decision maker's task is to choose the relation that, in one sense or another (depending on circumstances), is the average of the relations expressed by the experts. We shall now consider some types of relations, the ways of choosing the average of them, and their application to judging competitions.

4.2. Rankings. Let us suppose that n items are to be evaluated by N experts. Each expert has to state a relation expressed as a rank vector, indicating which of the items should be ranked number one, which should be ranked number two, and so on. To repeat, such a relation is called ranking. It is easy to see that the possible number of rankings equals $n!$. Let us denote the set of all possible rankings by R. We shall denote the ranking given by the ith expert by $\mathbf{r}_i = (r_{i1}, r_{i2}, \dots, r_{in})$, where r_{im} is the rank (place) of the mth item $(m = 1, \dots, n)$, as indicated by the ith expert.

For example, after an event involving four athletes—A, B, C, D—two judges have to arrange them in due order. Let us assume that the rankings they come up with are $\mathbf{r}_1 = (3, 1, 2, 4)$ and $\mathbf{r}_2 = (2, 1, 3, 4)$. This means that the first judge placed C first, A second, B third, and D fourth, and simultaneously, the second judge placed B first, A second, C third, and D fourth; or, $r_{11} = 3$, $r_{12} = 1$, $r_{13} = 2$, $r_{14} = 4$, $r_{21} = 2$, $r_{22} = 1$, $r_{23} = 3$, and $r_{24} = 4$.

Before demonstrating how the average estimator is chosen, we shall introduce the notion of distance (metric), $d(\mathbf{r}_i, \mathbf{r}_j)$, between rankings. Naturally, we would expect it, like all distance, to satisfy the following requirements (the reader will recall his elementary geometry):

1. The distance between rankings cannot be negative: $d(\mathbf{r}_i, \mathbf{r}_j) \geq 0$; a distance equals zero if and only if the rankings are identical: $d(\mathbf{r}_i, \mathbf{r}_j) = 0$ implies $\mathbf{r}_i = \mathbf{r}_j$, and vice versa.

2. The distance is symmetric: $d(\mathbf{r}_i, \mathbf{r}_j) = d(\mathbf{r}_j, \mathbf{r}_i)$.

3. The distance satisfies the triangle inequality $d(\mathbf{r}_i, \mathbf{r}_j) \leq d(\mathbf{r}_i, \mathbf{r}_k) + d(\mathbf{r}_k, \mathbf{r}_j)$.

In view of the requirements of examinations by experts, these three axioms have been supplemented by the following:

4. Let the rankings $\mathbf{r}'_i$ and $\mathbf{r}'_j$ be obtained from the rankings $\mathbf{r}_i$ and $\mathbf{r}_j$, respectively, as a result of a redesignation of the objects; then

$$d(\mathbf{r}'_i, \mathbf{r}'_j) = d(\mathbf{r}_i, \mathbf{r}_j).$$

5. The minimum positive distance between rankings equals 1.

6. If the ranking $\mathbf{r}_k$ lies between the rankings, $\mathbf{r}_i$ and $\mathbf{r}_j$, then

$$d(\mathbf{r}_i, \mathbf{r}_j) = d(\mathbf{r}_i, \mathbf{r}_k) + d(\mathbf{r}_k, \mathbf{r}_j).$$

It can be proved (see [15, 29]) that only one metric satisfies all six requirements. In this metric, the distance between the rankings $\mathbf{r}_i$ and $\mathbf{r}_j$ is determined by the formula

$$d(\mathbf{r}_i, \mathbf{r}_j) = \tfrac{1}{2} \sum_{m=1}^{n} |r_{im} - r_{jm}|.$$

In our example with athletes A, B, C, D and two judges, the distance between the rankings is

$$d(\mathbf{r}_1, \mathbf{r}_2) = \tfrac{1}{2}[|3-2| + |1-1| + |2-3| + |4-4|] = 1.$$

A ranking $\tilde{\mathbf{r}}$ such that the sum of distances from it to the initial ones is the least of all possible variables is often chosen as the average of several rankings $\mathbf{r}_1, \dots, \mathbf{r}_N$. Such a ranking $\tilde{\mathbf{r}}$ is called the *Kemeni median.* The Kemeni median, $\tilde{\mathbf{r}}$, is the solution of the minimization problem of the sum of distances from $\mathbf{r}$ to $\mathbf{r}_1, \dots, \mathbf{r}_N$, over all rankings $\mathbf{r}$ from the set R, or is determined by the formula

$$\sum_{k=1}^{N} d(\tilde{\mathbf{r}}, \mathbf{r}_k) = \min_{\mathbf{r} \in R} \sum_{k=1}^{N} d(\mathbf{r}_k, \mathbf{r}).$$

The solution of this problem in the general case is very complex, and we cannot dwell on it here (see [**15**]).

Now we shall examine different principles of choosing the resulting relation in the ranking case and take a look at how they developed.

4.3. Shortcomings of the majority principle. One of the best known and perhaps oldest principles of choosing the resulting relation is the *majority* *principle*. This is what it is about. Suppose we have N rankings of m objects: $\mathbf{r}_1, \ldots, \mathbf{r}_N$. In this case, first place in the resulting ranking goes to the object placed first by the majority of the experts; second place goes to the object placed second by the majority of the experts; and so on.

Let three rankings be given to four items:

$$\mathbf{r}_1 = \begin{pmatrix} 1 \\ 2 \\ 3 \\ 4 \end{pmatrix}, \quad \mathbf{r}_2 = \begin{pmatrix} 1 \\ 2 \\ 4 \\ 3 \end{pmatrix}, \quad \mathbf{r}_3 = \begin{pmatrix} 1 \\ 2 \\ 4 \\ 3 \end{pmatrix}.$$

The resulting ranking will be as follows:

$$\tilde{\mathbf{r}} = \begin{pmatrix} 1 \\ 2 \\ 4 \\ 3 \end{pmatrix},$$

because the majority assigned first place to the first object, second place to the second object, third place to the fourth object, and fourth place to the third object.

Nevertheless even early research, by Jean-Antoine Condorcet (1743–1794) and Jean-Charles Borda (1733–1799),[13] into the problem of choosing the best possible ranking (of objects), noted the insufficiency of the procedure for determining the resulting relation on the majority principle.

Here is an example (cf. **[15]**) illustrating Condorcet's point of view. Let us assume there are twenty objects (alternatives) a_i $(i = 1, \ldots, 20)$ and their rankings given by ten experts are as follows.

$$\begin{array}{cccccccc}
\mathbf{r}_1 & \mathbf{r}_2 & \mathbf{r}_3 & \mathbf{r}_4 & \mathbf{r}_5 & \mathbf{r}_6 & \cdots & \mathbf{r}_{10} \\
a_2 & a_2 & a_1 & a_3 & a_4 & a_5 & \cdots & a_9 \\
a_1 & a_1 & a_3 & a_1 & a_1 & a_1 & \cdots & a_1 \\
a_3 & a_3 & a_4 & a_4 & a_3 & a_3 & \cdots & a_3 \\
a_4 & a_4 & a_5 & a_5 & a_5 & a_4 & \cdots & a_4 \\
\cdots & \cdots & \cdots & \cdots & \cdots & \cdots & \cdots & \cdots \\
a_{19} & a_{19} & a_{20} & a_{20} & a_{20} & a_{20} & \cdots & a_{20} \\
a_{20} & a_{20} & a_2 & a_2 & a_2 & a_2 & \cdots & a_2
\end{array}$$

To make it more graphic, the alternatives a_i are located according to the places held in each ranking. Under the majority principle, a_2 in our example has to be recognized as the best, though it is obvious that there is more reason to consider it the worst.

Sometimes additional requirements are imposed to eliminate this shortcoming. Thus, an alternative may be declared to be the best if at least half of the experts consider it so. However, in real examinations this is not often the case. Condorcet suggested the following principle for choosing the best ranking. Based on the rankings obtained from the experts for each pair of

[13] Philosopher and mathematician Condorcet wrote his *Sketch for a historical picture of the progress of the human mind* in which he predicted the possibility of the prolongation of man's life.

alternatives a_i, a_j, they compute S_{ij}—the number of the experts who think that a_i is better than a_j. Accordingly, S_{ji} is also found. If $S_{ij} > S_{ji}$, then the alternative a_i is considered to be preferable to a_j ("a_i is preferred over a_j" is written $a_i \succ a_j$). The alternative a_i is declared to be *the best alternative—the Condorcet alternative,* if $S_{ij} > S_{ji}$ for all $j \neq i$. In the foregoing example, this alternative is a_1.

Nevertheless, using the Condorcet principle of choice may entail a paradox he himself pointed out, which follows from the intransitivity of collective preferences. Intransitivity means violation of the law of transitivity, i.e., when $a_1 \succ a_2$ and $a_2 \succ a_3$, still $a_3 \succ a_1$.

Let us cite an example to illustrate the Condorcet paradox. Assume that three experts ranked three alternatives:

$$\mathbf{r}_1 = \begin{pmatrix} a_1 \\ a_2 \\ a_3 \end{pmatrix}, \quad \mathbf{r}_2 = \begin{pmatrix} a_3 \\ a_1 \\ a_2 \end{pmatrix}, \quad \mathbf{r}_3 = \begin{pmatrix} a_2 \\ a_3 \\ a_1 \end{pmatrix}.$$

In this case $S_{12} > S_{21}$, $S_{23} > S_{32}$, but $S_{13} < S_{31}$; hence, $a_1 \succ a_2$, $a_2 \succ a_3$, but $a_3 \succ a_1$, and therefore the Condorcet alternative does not exist in this case.

Note that the number of examinations resulting in the Condorcet paradox is, on the average, slightly less than $\frac{1}{10}$ of their total number. In real-world problems, when the experts' opinions may diverge a great deal, the paradox occurs with an even higher probability than $\frac{1}{10}$.

Another improvement of the majority principle was suggested by Borda. The objects ranked by an expert are "weighted": the last is given a weight of zero; the next-to-the last, a weight of one, and so on. If s_i is the total of the weights assigned to the alternative a_i by all the experts, the resulting relation is declared to be $\tilde{\mathbf{r}} = (a_{i1}, a_{i2}, \dots, a_{in})$ for which $s_{i1} \geq s_{i2} \geq \cdots \geq s_{in}$. Borda's method also has its shortcomings. For example, the Condorcet alternative—i.e., the one better than any other in paired comparison—may fail to be chosen as the best by the Borda principle.

Consider a simple example. Let us assume that five experts have ranked the objects $a_1, a_2, \dots, a_5$ as follows:

$$\mathbf{r}_1 = \begin{pmatrix} a_1 \\ a_3 \\ a_2 \\ a_5 \\ a_4 \end{pmatrix}, \quad \mathbf{r}_2 = \begin{pmatrix} a_1 \\ a_2 \\ a_4 \\ a_3 \\ a_5 \end{pmatrix}, \quad \mathbf{r}_3 = \begin{pmatrix} a_1 \\ a_2 \\ a_5 \\ a_3 \\ a_4 \end{pmatrix},$$

$$\mathbf{r}_4 = \begin{pmatrix} a_2 \\ a_3 \\ a_1 \\ a_5 \\ a_4 \end{pmatrix}, \quad \mathbf{r}_5 = \begin{pmatrix} a_2 \\ a_4 \\ a_3 \\ a_1 \\ a_5 \end{pmatrix}.$$

The best, according to the Condorcet principle of paired comparisons, is the object a_1 (as the reader can verify). Yet, according to the Borda principle, it is worse than a_2, since $s_1 = 15$ and $s_2 = 16$.

These shortcomings are absent from the selection principle suggested by Kemeni. As was mentioned, this principle is based on the choice of the average relation (ranking) least removed from those expressed by the experts. Below we discuss our own methods for finding the average in figure skating.

4.4. Judging figure skating. Let us start with singles skating, which consists of compulsory exercises (the school figures), a short program, and the free skating program.[14]

The compulsory program includes three compulsory figures out of a larger number of figures recognized in ISU (the International Skating Union) competitions, Olympic Games, and USSR competitions. Since 1980, lists of school figures have been published by ISU for each season. Each figure (the loop, bracket, paragraph, etc.) is evaluated on a six-point scale, from zero for omission to six for faultless execution. The mark is given for correct tracings, reasonable speed, graceful movements, and so on.

The short program (introduced in 1971, to give added weight to the optional parts of a competition) is comprized of a group of required elements (jumps, spins, serpentines, etc.). In 1975, four groups of required elements were introduced, one of which is drawn for the coming season. This group is excluded from the draw the next year, and so on, so that all the four groups are skated over the four-year interval between Olympics.

The short program is performed to music of the skater's choice and comprises seven required elements linked by step sequences. The program must not include any additional elements; its presentation is limited to two minutes.

The short program is given two marks, for *required elements* and *presentation.*

Judging in the USSR is somewhat different from international judging. At competitions held in the USSR, the judges keep count of the errors in each required element, taking out from 0.1 for a small error up to a whole point for the omission of a required element. Then the deductions are subtracted from the maximum six mark. The remainder is the mark given for a required element.

The second mark (also a maximum six) may notably differ from the first. It is given for the general impression made by the composition of the program, artistic quality, harmony of music and motion, and so on.

In international judging, points are not deducted from the maximum six mark, but from a basic mark not exceeding six. The basic mark is determined by each judge individually for each contestant, depending on the latter's ability, skating skill, and other considerations that may be purely subjective. The basic mark may be lowered only for some, rather than for all, errors (the interested reader is referred to the literature, e.g., [**13**]).

The free skating program can be any arrangement of moves by the skater

[14] Recently, the requirement of compulsory exercises was dropped, and only the short and free skating programs remain. (Editor's note)

(jumps, spins, spirals, step sequences, etc.) performed to music of the skater's choice. In the free skating program the skaters seek to demonstrate their ability, artistry, good taste in music, and their personalities. The time allotted for the free skating program is four minutes for women and four and a half for men. It consists of several parts differing in tempo and elements, yet harmoniously combined. Many programs performed by top-scoring skaters consist of more than twenty jumps, up to five spins, several step sequences, and other elements. While it affords the skater great opportunities, the free skating program is subject to strict rules, which provide for efficient judging (see [13]).

The free skating program is also given two marks, for *technical merit* and *artistic impression.* The first mark is given according to the difficulty of the program and accuracy and sureness of execution. The second mark reflects the originality, harmonious composition and artistic merits of the program, utilization of the ice surface, and so on.

Though qualitatively different, these marks are largely interdependent. For example, it is easier for a skater whose program contains no complex elements to obtain a higher mark for artistic impression. Both marks are given out of the maximum of six. The judges appraise the free skating program by sight, without quantitatively comparing the skaters as to how fast they skate, how high they jump, or how long they spin. Besides, for some new elements no generally accepted yardsticks exist at all. In such cases of particular importance is the judges' experience, lack of bias, good taste, and so on.

At the end of the three-part competition, the skater's place has to be decided. It used to depend on the sum of the points scored in all three parts of the competition. Subsequently the system of judging was improved. In 1980, an ISU congress adopted a new system for calculating the final results, whereby the points scored by a skater in one part of the competition determine his place in that part alone.

Now we are ready to discuss judging and scoring in singles skating competitions. In figure skating an open system of marking is used, i.e., at a signal from the referee (the head judge) the judges show their marks (or they may be flashed on the scoreboard).

To help them arrive at a common standard, the judges are given preliminary information on the contestants' abilities and the requirements set for the competition. The marks given to the first contestant, whether for the school figures, the short, or the free skating programs, are, in a sense, discussed collectively. First the referee and his assistant collect from the judges the marks they gave to the first contestant. After calculating the average, the referee lets the judges know the panel's highest and lowest marks and the average, upon which any judge may adjust his mark within that range.

The contestants are given one mark for each of the three compulsory figures, which are entered in a card provided for each skater. Then each judge sums up the points he gave each skater for all three figures, placing the skater with the highest points first, the next one second, and so on, down the line. If he gave the same number of points to two or more contestants, he puts them in the same (better) position. For example, two or three (or more) contestants with the same score claim first, second, and third places. Each of them has the better (first) place entered in his card. Before defining a skater's final standing in the compulsory figures, the referee finds out how each judge has ordered (placed) the contestants.

The final assignment of places has a number of special features.

A panel of judges may consist of three, five, seven, nine, or eleven members. Under the rules, a skater's place is decided by a majority vote. For example, the nine judges placed skater A 1–1–2–2–1–1–1–1–1; seven judges put him first and two put him second. Skater B's placings are 2–2–1–1–2–2–2–2–2; he has only two first places and seven second places. So, finally, A gets first place and B gets second place.

Here is another example. A has been placed 3–3–4–4–3–3–4–2–4. He has no chance of winning second place; the one second place he has got is counted as third in calculating the number of third places. Thus he has five third places (a majority, the number of judges being nine) and four fourth places. B, whose placings are 4–4–3–3–4–4–3–4–3, is one vote short and so loses third place to A.

The place a skater is put in by the majority of the judges is his *deciding place.* A clear definition of the deciding place can be given with the help of mathematical induction.

A skater's deciding place is first, if an absolute majority of the judges (at least three out of five, four out of seven, five out of nine, and so on) have put him first. For a skater, kth is his deciding place if no higher place happens to be the deciding one for him, and if an absolute majority of the judges have put him in the kth or a higher place.

For example, if the places given to a skater by seven judges are 2–1–1–3–2–1–1, then his deciding place is the first. Suppose the places the same judges have given another skater are 1–4–3–2–1–4–3. In that case, third is the deciding place. Indeed, this skater was placed first by only two judges, or less than an absolute majority of four. He was placed first and second by three judges or, again, less than an absolute majority. Therefore, second is not his deciding place either. But he was placed third, second, or first by five of the seven judges, and that is an absolute majority. Hence, his deciding place is the third.

If two or more contestants have the same deciding place, then the skater who has more deciding places is put ahead of others. Therefore, when computing the results, they compute, besides the deciding place (DP), also the *number of deciding places* (NDP), the higher places being regarded as deciding.

For example, a panel of five judges placed a skater 1–2–3–3–4. The skater's deciding place is third (DP=3), and his number of deciding places is four (NDP=4). Another skater's places are 2–1–4–4–3. His DP=3 and NDP=3. The latter will score fourth, and the former, third.

Also, two or more skaters may happen to have the same number of deciding places. Therefore, besides computing the deciding places and their number, the *sum of deciding places* and the *sum of places* also have to be found. If the DPs and NDPs are equal, a skater's place is determined by the sum of deciding places (SDP), and if that too is the same, the better place will go to the skater with the smaller sum of places (SP). For example, contestant A has been placed 1–2–3–3–4. His DP=3, NDP=4, SDP=9, SP=13. Contestant B has been placed 1–1–3–3–5. His DP=3, NDP=4, SDP=8, SP=13. Accordingly, B scores third and A scores fourth.

Another example: A has been placed 1–1–2–2–3. His DP=2, NDP=4, SDP=6, SP=9. B has 2–2–1–1–4. His DP=2, NDP=4, SDP=6, SP=10. A emerges in second place, and B wins third place.

Should several contestants have the same deciding places, the same number and sum of deciding places, and the same sum of places, then these contestants share the final place. That is how final places are assigned in the compulsory program.

The final result of the competition is calculated from the marks won for each section, with a *weight factor* assigned to each of the three marks. Under the new ISU rules the weight factor for the school figures is 0.6. It means that the place won by a skater for the compulsory figures is multiplied by this factor. For example, a skater who scores first in the compulsory figures has the factored place 0.6; the factored place of the runner-up is 1.2; the one in the third place has 1.8, and so on. The marks in the short and free skating programs have weight factors of 0.4 and 1, respectively.

Results in the free skating program are calculated by adding the two marks (for technical merit and artistic impression) given by each judge. If two or more contestants are given the same number of points by a judge, the deciding mark is that for technical merit. For example, a judge gives 5.7 and 5.8 to one skater and 5.8 and 5.7 to another. Both have an equal number of points, 11.5. Nevertheless, the higher place goes to the skater with a higher mark for technical merit. If, however, both have the same mark for technical merit, then they share the (better) place.

The final results in the short and free skating programs are determined in the same way as in the compulsory program. That is, after each contestant has been ranked by the judges, the head judge calculates the final result from the majority of the skater's better places. For this, he finds the deciding place (DP), the number of deciding places (NDP), the sum of deciding places (SDP), and the sum of places (SP). Recall that the weight factor for the short program is 0.4 and for the free skating program is 1. It means that the place given to a skater in the short program is multiplied by 0.4 and in the free skating program by 1.

To determine a skater's place by the results of only the short and free skating programs, the factored places, obtained by multiplying the skater's place by the respective factor, are added together. The skater with a smaller total gets a better place. If two skaters have equal factored places, the better place goes to the skater who holds the higher place in the free skating program.

To determine a contestant's place in a three-part competition, his factored place in the compulsory figures is added to his places in the other two sections.

For example, assume that the skaters A, B, and C win the following places:

	1. Compulsory	2. Short	3. Free Skating
A	1	3	3
B	2	2	1
C	3	1	2

The summarized figures are:

$$s(A) = 1 \cdot 0.6 + 3 \cdot 0.4 + 3 \cdot 1.0 = 4,8,$$
$$s(B) = 2 \cdot 0.6 + 2 \cdot 0.4 + 1 \cdot 1.0 = 3,$$
$$s(C) = 3 \cdot 0.6 + 1 \cdot 0.4 + 2 \cdot 1.0 = 4,2.$$

Since $s(B) < s(C) < s(A)$, first place goes to the skater B, second to C, and third to A.

In conclusion, we shall touch briefly on the judging of pair skating and dance.

Pair skating competitions consist of a short and a free skating program. The elements included in the short program (jumps, spins, lifts, step combinations, etc.) are divided into four groups of seven required elements each. One of these groups, chosen by a draw, is performed in the course of a season at all major competitions. The elements not in this group cannot be included in the short program, which is two and a half minutes long. Any elements performed beyond this time limit are not marked. The short program is given two marks, for the required elements and for presentation. Each pair chooses its own music, order of the required elements, and transitions.

The highest mark that can be given for the short program is six points. Under USSR competition rules, this mark can be lowered 0.1 of a point for small errors, 0.2–0.3 of a point for insignificant errors, 0.4–0.6 of a point for significant errors, and up to 0.9 of a point for serious errors. The marks for required elements and presentation are added together. Judging of the short program for pairs is the same as for singles.

The five-minute free skating program is also given two marks, for *technical merit* and *composition and execution.* The first mark refers to technique, perfection, and sureness in executing the elements of the program (a top-scoring skater's routine may contain more than twenty elements). The second mark is given for composition, the choice of music, the style and beauty of presentation, and synchronized action. The sum of these two marks makes up the total points scored in the free skating program. If two or more pairs have been given the same number of points by a judge, the pair with a higher mark for technical merit gets a better place. A pair's final score for the free skating program is decided in the same way as in the singles. The final result in pair skating is determined from the sum of two final marks, the mark (place) for the short program having the weight factor 0.4, and the mark (place) for the free skating program having the weight factor 1.

Ice dancing competitions consist of three sections: compulsory dances, original set pattern dance, and free dancing.

The fifteen compulsory dances executed at international competitions (fox-trot, six varieties of the waltz, three varieties of the tango, rhumba, blues, march, quickstep, paso doble) are divided into four groups. One of the groups is chosen by a draw on the eve of the competition. Each couple skates three compulsory dances, also chosen by a draw, and one original set dance. The compulsory dances have designated diagrams, timing, steps, and music. The rhythm of the original set dance is fixed every two years by the ISU Ice Dance Committee.

Each compulsory dance is marked from zero to six. The original set dance is given two marks—also out of six—for *composition* and *presentation.*

Free dancing, of four minutes' duration, is of a distinct athletic-choreographic character and has elements different to pair skating. The skaters choose their own music (being limited to three changes of tune), rhythm, tempo, and style of dancing.

Free dancing is marked separately for composition and presentation. The final marks for each of the three sections of the competition are found by the method we already know. The winners are determined by the overall mark calculated from the marks (places) for compulsory dances, the original set dance, and free dancing, with their respective weight factors of 0.6, 0.4, and 1.

Thus, we have now been introduced to the principles of judging figure skating. It is easy to see that it is a modified Condorcet principle, with some elements of the Borda method as far as the singles are concerned, where the Condorcet principle cannot determine the winner.

It is natural, therefore, that contemporary judging should share some of the shortcomings of the Condorcet and Borda methods. It becomes especially obvious when widely divergent marks are given, for instance, by biased judges.

In the next section we shall show that modern mathematics offers methods for obtaining fairly objective final marks, even in the face of bias.

4.5. Multiround examinations by experts and their modeling. Here we shall discuss two ways of offsetting experts' bias. One of them is to stage examinations in several rounds, asking the same question several times. Moreover, the experts may or may not be given information about the previous rounds—for example, about other experts' opinions or about the general average estimator. Of course, if such additional information is provided, the experts are able to adjust their conclusions from round to round.

The other way is to model (rather than actually conduct) a multiround examination, using experts' opinions stated just once, in order to adjust their average opinion **[15, 16]**.

Let us take a closer look, at this method. Let R as before, be the set of possible relations, with the metric $d(\mathbf{r}_i, \mathbf{r}_j)$ (see 4.2). Assume that the panel of experts has n members. Let $\mathbf{r}_i^0$ be the ith expert's "objective" opinion

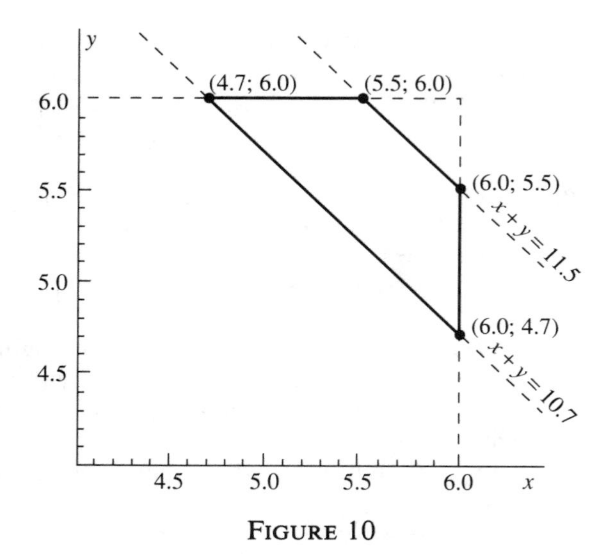

FIGURE 10

and $\mathbf{r}_i^*$, his "subjective" (opportunistic) opinion. That two such different opinions can exist is practically indisputable. For example, any sports judge forms for himself an "objective" estimate of an athlete's ability, on the one hand, while on the other, as a representative of his country, member of a certain school (trend), and representative of a certain sports society, he cannot dismiss his "subjective" (somewhat partial) approach.

It is natural to suppose that the ith expert always states a relation somewhere between $\mathbf{r}_i^0$ and $\mathbf{r}_i^*$.

Consider the following interesting and important example. A judge gives a skater two marks, for technical merit and for artistic impression. The place this judge gives to the skater is determined by the sum of these two marks. Assume that the relation $\mathbf{r}^0 = (5.7; 5.8)$ expresses the judge's "objective" and the relation $\mathbf{r}^* = (5.4; 5.3)$ his "subjective" opinions. Pairs of marks whose sum is more than 10.7 and less than 11.5 form a set of relations lying between $\mathbf{r}^0$ and $\mathbf{r}^*$. This point set is shown in Figure 10 and is defined by the following constraints–inequalities:

$$x + y \leq 11.5, \qquad x + y \geq 10.7,$$
$$0 \leq x \leq 6, \qquad 0 \leq y \leq 6,$$

where x is the mark for technical merit and y is that for artistic impression.

The geometry of the relation set situated between two given relations can be highly varied. For instance, if each judge gives just one mark, as they do in gymnastics, then we have to do with an ordinary interval of the number axis.

Suppose the experts provide n relations $\mathbf{r}_1, \mathbf{r}_2, \dots, \mathbf{r}_n$. Having solved the problem of finding the Kemeni median (see 4.2), we find the mean $\tilde{\mathbf{r}}_1$ of the relations provided by the experts. Next we find the distances $d(\mathbf{r}_i, \tilde{\mathbf{r}}_1)$ from this mean to the relation $\mathbf{r}_i$ for all $i = 1, \dots, n$. Lastly, we introduce conditional coefficients of judges' objectivity $\alpha_1, \alpha_2, \dots, \alpha_n$ by the formulas

$$\alpha_i = 1 - \frac{d(\mathbf{r}_i, \tilde{\mathbf{r}}_1)}{\sum_{i=1}^n d(\mathbf{r}_i, \tilde{\mathbf{r}}_1)}.$$

Thus, the closer $\mathbf{r}_i$ is to the mean $\tilde{\mathbf{r}}_1$, the greater the objectivity coefficient. Allowing for the objectivity coeffcients, we find a new mean relation $\tilde{\mathbf{r}}_2$ that minimizes the sum of the distances to the relations $\alpha_1\mathbf{r}_1, \alpha_2\mathbf{r}_2, \dots, \alpha_n\mathbf{r}_n$, repeat this operation, and so on. In the limit we shall obtain the relation $\mathbf{r}$, which precisely should be chosen as the mean.

In practice it is often enough to find the relation $\tilde{\mathbf{r}}_2$, as even that is already close to the true Kemeni median of the "objective" judgements $\mathbf{r}_1^0, \dots, \mathbf{r}_n^0$.

In the light of the foregoing, it would be interesting to discuss judging in gymnastics and diving. In these sports, the final results are calculated disregarding the lowest and highest marks, which favors objective judging. Apart from that, in gymnastics the head judge or the panel can change an athlete's mark for an exercise if the judges' opinions sharply diverge. Let us now examine mathematical models of judging in gymnastics.

4.6. Hierarchic examination (judging competitive gymnastics). Men's competitions in gymnastics consist of six kinds of exercises—the high bar, parallel bars, pommel horse, rings, the floor exercises, and the vault (over the length of the horse). Women compete in exercises of four kinds: the beam, asymmetric bars, floor exercises, and the vault (over the width of the horse).

At Olympics and world championships three types of competitions are held:

(a) the team event, with compulsory and voluntary programs in each of the six exercises—the high bar, parallel bars, rings, the pommel horse, the vault, and the floor exercises. This competition is a preliminary for the individual finals in each of the six exercises. Of the six members of each team, the results of the five best gymnasts on individual apparatus are credited to the team;

(b) individual finals, with a voluntary program on six apparatus. The 36 gymnasts who made the best showing in the team event are selected for this competition. The competitors have 50% of the total points scored in the team event added to their score.

(c) finals on individual apparatus, for which the six gymnasts who had the best scores on each apparatus in the compulsory and voluntary programs of the term event are selected. They have 50% of their points scored in the team event added to their score.

The starting mark for performance on each apparatus is 9.4 points, to which are added bonus decimal points for risk, originality, and virtuosity.

A country may be represented at Olympics and world championships by a maximum of three gymnasts in individual finals and two in apparatus finals.

Each exercise is marked out of ten, with decimals and hundredths. Thus, the starting mark of 9.4 for an exercise can be increased by up to 0.2 for risk and difficulty, by up to 0.2 for originality, and by up to 0.2 for virtuosity.

The judges follow the FIG (Fédération Internationale Gymnastique) Codes of Points for men and women, based on the analysis and generalization of experience accumulated at numerous competitions, combined with special research. The Codes are adopted by the FIG Technical Committees. FIG was founded in 1881. It sponsors world and European championships, as well as competitions within the Olympic Games framework, which are held every other year.

In a gymnastics competition, the winner should score the highest total of points, whether in the all-around or the apparatus final. Thus, the basis of the total is formed by the points received on individual apparatus. At FIG competitions, Olympics, international, and continentwide competitions, the panel of judges consists of five members, one superior judge and four international judges, representing the participating national federations. The panel of judges for apparatus final consists of two superior and four other judges. A superior judge must have ample specialized knowledge, excellent ability, and be free from bias. He is in full charge of the panel of judges on his apparatus. The superior judge must objectively evaluate all exercises, supervise the panel's work, check the scores, and intervene if a judge should be biased. His score, added to the average of the two middle scores, given by the four judges and divided by two, is the final base score that is used if a score is disputed or has to be adjusted.

Obviously enough, a superior judge is selected from among the most reliable and unbiased judges, those in particular who enjoy the trust of the Gymnastics Federation.

Apparatus exercises are marked out of ten with decimals. The final score is the arithmetic mean of the two middle scores of the four judges. For example, if the scores are 9.5–9.6–9.7–9.8, the lowest and highest scores of 9.5 and 9.8 are discarded, and the average score is (9.6+9.7)/2=9.65. The difference between the two middle scores should not exceed 0.1 at the average score of 9.6 to 10; 0.2 at the average score of 9 to 9.55; 0.3 at the average score of 8 to 8.95; 0.5 at the average score of 6.5 to 7.95; 0.8 at the average of 4 to 6.45; and 1 in all other cases. For example, a score of 9.5–9.6–9.8–9.8 is inadmissible since the average is 9.7 and the difference $9.8 - 9.6 = 0.2 > 0.1$. In such cases, the superior judge calls a conference and asks the judges to revise their marks. Only the final score is flashed.

Thus, judging in competitive gymnastics is on two levels. First the scores are given on the lower level (the four judges). If the scores differ a great deal, the superior judge calls the panel together to view a video of the routine and adjust their scores and to explain why they deducted points or did not count errors.

Naturally, such a hierarchic organization of judging, with the two extreme marks disregarded, permits more objective evaluation in such a qualitatively complex sport as gymnastics. Note that the impact of one biased judge in gymnastics is considerably less than in figure skating, even though the number of judges there is greater. Two biased judges can seriously affect a gymnastics score too. However, then the difference between the middle scores becomes more consistent, and the superior judge takes part in working out an unbiased evaluation. In figure skating, a mark given by a judge is final. So, if the majority of the panel belongs to one school of figure skating, this fact can adversely affect the scoring of the skaters belonging to a different school.

4.7. Examination by experts. An overview. One notable feature of the progress of science and technology today is their mathematicization and computerization. Mathematics is advancing everywhere, invading every area of human knowledge and practice. The list of traditional fields where mathematics is applied—physics, chemistry, mechanics, engineering—has now been joined by economics, sociology, biology, medicine, criminalistics, psychology, linguistics, and so on. Presumably, sooner or later this list will include still other areas of knowledge.

As the reader can see, sport too has been invaded by mathematics.

What, then, are the characteristics of the mathematical methods that penetrate into the humanities and sport? First of all, it is clear formulation of the problems, exact definitions, formalized (logically valid) reasoning, utilization of mathematical techniques, and quantitative estimation of phenomena. Conversely, the humanities widely resort to descriptive reasoning, loose terminology, and polemics and appeals to common sense, feeling, and analogy.

Penetration of mathematics into new domains is accompanied by a reciprocal process. As it deals with sociological problems, specific life situations, processing of information, economic decisions, decision making under conditions of uncertainty, and other problems, mathematics absorbs and methods common to the humanities. Once sharp, the dividing lines between these sciences are increasingly blurred.

"Indeed, let us ask ourselves where it has sprung from, this difference between the methodologies of the exact sciences and the humanities. Why did the formal mathematical apparatus come to be applied in the exact sciences very early on and only recently—and merely as an auxiliary—in the humanities? Is it perhaps that scientists are smarter than scholar? Certainly not! Simply the phenomena making up the subject matter of the humanities are immeasurably more complex than those in the exact sciences. They are much harder, if not impossible, to formalize. Each of them depends on a much wider range of causes. Ironically, the verbal (descriptive) method of investigation is more accurate here than the formal logical approach. And yet we are sometimes simply obliged to construct mathematical models here too. If not exact, then approximate models. If not to get a simple answer to the question, then to get some understanding of the thing." [15]

Practical problems often make it necessary to reconcile different points of view on obtaining a solution that would satisfy different, sometimes inconsistent, requirements or criteria. For example, the management of an industrial enterprise seeks to maximize output and profits and minimize costs. An

[15] I. Grekoka, *Methodological features of applied mathematics at the modern stage*, Problems in Philosophy, No. 6, 1976, p. 107. (Russian)

antiaircraft defense system has to be designed to provide for maximum efficiency at a minimum expenditure.

In such trade-off problems, as they are known, one has to adjust different requirements so as to find a reasonable compromise. Naturally mathematics has methods for finding compromise solutions to trade-off problems. These methods, however, are far from perfect.

"So far, practially the only thing capable of producing a quick and successful compromise solution is man's reason, his 'common sense'. Man to this day is the superb master of compromise, and without his participation no solution—not optimal, perhaps, on any one of the criteria, but feasible on all of them together—can as yet be chosen in a trade-off problem. Mathematics, in its present form, can only use the notions 'more', 'less', and 'equal', but not 'feasible', 'practically equivalent', or similar notions typical of human thinking. Evidently not every 'better or worse' can be reduced to 'more or less' and if it can, we do not know how to do it. In making a decision, one simply takes in the situation at large and chooses an acceptable alternative. As for mathematics, its job in such cases is not to furnish the final decision but to help choose one. To provide the decision maker with a maximum of relevant information in an expressive, easily perceivable form, to show what consequences will result (on a number of criteria) from each of the possible alternative decisions, having discarded all noncompetitve ones first."[16]

Coaches of contending teams provide graphic examples of a search for compromise solutions drawing on experience, intuition, and so on.

Apparently, scientific method in the humanities is going to be based on mathematics interfaced with nonformal procedures: the peerless characteristics of the human mind, intuition, and experience, strenghtened by the ever greater possibilities afforded by computers.

Mathematics is the basis of formal methods for examining a suitation. Common sense and experience are the basis of formal methods which do not operate with formal, say numerical data, but with qualitative characteristics like "more beautiful", "more complex", "simpler", "more comfortable", and so on. These are just the kind of characteristics that judges have to deal with at sporting events.

Mankind has long been familiar with the method of joining formal and nonformal principles in a single procedure. This method is examination by experts. Experts' estimates provide a natural opportunity to apply accumulated human experience, mankind's social memory, to the big problems of social and economic control and the lesser problems of judging competitive sports.

We have already been introduced to some of the specific procedures involved in organizing examinations. Other procedures are also used—some of them simpler, others more complex. Some examinations are based on discussion, others rule it out. A case in point is a doctors' conference, with specialists talking over all the pros and cons of a diagnosis and usually reaching a coordinated group decision. The military conference at Fili, presided

[16] Ibid.

over by Fieldmarshal Kutuzov, as described by Leo Tolstoy in his novel *War and Peace*, was conducted in a different manner. There, each general present assessed the situation and made his suggestions. Having heard them all, Kutuzov pronounced his own final decision.

Very often examination procedures depend on tradition, the subject of examination, the possibility of processing the information, and the mathematical methods employed. Presumably the oldest body to make group decisions was the tribal council of the elders, which made decisions on all matters affecting the tribe.

The common procedure of food tasting, too, emerged in the remote past. Its object is to appraise the quality of foodstuffs and grade them according to certain standards. Like judging, tasting must result in a conclusion. And a conclusion is the result of consensus.

Repeated examinations make it possible—in addition to obtaining a coordinated decision—to evaluate the experts themselves. Experts can have "weights" assigned to them, the way we saw it done (see 4.5) to obtain an unbiased appraisal of a skater's performance. The less an expert's, judgement deviates (on the average) from the group average, the greater the weight that can be assigned to him. This provides feedback, which helps establish a common information base and achieve a better consensus of experts.

The measure of consensus can be estimated on different principles. Besides the already familiar principle of majority vote and its modifications, other principles can be used as well. Sometimes the measure of consensus is the standard deviation (variance) of experts' numerical estimates from the average estimator (for a definition of variance see 5.1). The less the variance, the greater the consensus; the greater the variance, the less consistent the experts' opinions, and the poorer the quality of the examination.

New methods have been developed for tackling political, economic, social, and other problems. One of them, widely known in the United States, is the Delphi method (so named in allusion to the legendary Delphi oracle of ancient Greece that predicted the future and answered knotty questions).

We shall describe the Delphi method following N. N. Moiseev [**4**]. It was developed in the mid-1950s by the Rand Corporation, an American "think tank", and is oriented at forecasting and estimating the probability of events. Its information processing techniques are not new.

Nevertheless, the procedures of working with experts are well worked out and are governed by a number of principles, the simplest of which are the following:

1. Only recognized experts in the field can be asked to participate in an examination.
2. Since any expert may be wrong, it will take the opinion of a sufficiently large number of them to get meaningful information on the problem.
3. The questions must be relatively simple and be formulated clearly enough to rule out misinterpretation.
4. Examinations should be supervised by permanently functioning groups of investigators, since only well-formulated questionnaires (which should be

developed by trained specialists) can be successful in a comparatively involved examination.

The Delphi method allows no open discussion among the experts to prevent them from being influenced by psychological and emotional factors inevitable in a discussion. It provides for a procedure whereby the experts can request, obtain, and analyze additional information, including other experts' opinions. A provision is made for repeated examinations and additional questioning; the investigators' behavior is strictly regulated to prevent the experts from being influenced even indirectly.

All these measures are aimed at reaching a better consensus among the experts. However, if repeated questioning fails to achieve sufficient consensus on a problem (one that is uncertain or is the first of its kind), then this problem is considered unsolvable at the current level of knowledge.

The measure of consensus among the experts is the standard deviation or the "quartile". The value of the quartile is set before the examination. If, as a result of the examination, the value of the quartle turns out to be greater than the present value, another round is arranged, preceded by the experts' work with additional information.

The Delphi method was used to find solutions for many military problems and to forecast the probable state of the world in 1984, 2000, and 2100. For example, it was forecast that in 1984 the population of the globe would amount to 3.4 billion; [17] that water desalination would become common in farming; that a permanent station would be set up on the moon; that thanks to efficient control, the birth rate would continue to go down, and so on.

The basic idea underlying the method of prediction by experts is that group opinion is better than individual opinion. But even though it is mostly so, history knows situations in which this hypothesis failed. It is sufficient to recall the destiny of many great discoveries (the Copernican system) or the persecution of exponents of nontraditional views (Jordano Bruno, Galileo).

Much is being done in this country and abroad to improve the methods and organization of prediction by experts. A new approach was suggested in 1960 by G. S. Pospelov. The method was used to solve the problem of allocation of the funds for fundamental research in the natural sciences. The idea is to break down problems successively into simpler ones, questioning the experts on each specalized problem, processing the information, breaking down the problems already analyzed into more narrowly specialized ones, and so on.

Description of this system of integrated examinations by experts, as well as of the methods proposed by V. M. Glushkov and Yu. I. Zhuravlev, exceeds the scope of this book, and we refer the reader, for example, to **[4, 29]**.

---

[17] According to the UN Population Committee, in 1984 the population of the globe was 4.8 billion and was expected to reach 6.1 billion at the start of the next century.

5

Records! Records!

Mathematical methods of selecting, describing, and analyzing experimental data obtained from the observation of numerous random phenomena are the subject of *mathematical statistics*. Depending on the nature of the question and the available experimental material, problems in mathematical statistics may vary in form. Descriptions of some typical problems can be found in [**3**, **26**].

In recent decades, mathematical statistics has learned how to process observations containing not only quantitative but also qualitative information. There has sprung a new branch of statistics, known as *nonnumerical statistics*.

What is more, statistical processing of input information has generated a new discipline, now known as *applied mathematical statistics* (compare this with applied mathematics). Applied mathematical statistics develops and systematizes concepts, techniques, and mathematical methods and models intended for organizing the collection, standard notation, and classification and processing (by computers, among other means) of statistical data in order to ensure their convenient presentation and interpretation and to draw scientific and practical conclusions from them [**26**].

And now, let us see how mathematical statistics can be used to forecast athletes' achievements and records.

5.1. Random variables. Let us recall the concept of a random variable and its characteristics that we shall need for further discussion.

We shall apply the designation *random variable* to a variable that can, as a result of an experiment, take on a value unpredictably.

One example of a random variable is the number of hits scored by a marksman firing 60 shots. All we know in advance is what values this variable can assume, but we cannot possibly know how many times the marksman will hit the target. Another example is the number that will face up at the roll of a die. Random variables that take only isolated values from some finite or denumerable set are called *discrete random variables*.

The concept of random variable plays an important role in modern probability theory. Further on we shall denote random variables by capital letters and their possible values by small letters. Assume that a random variable X can take on the values $x_1, x_2, \dots, x_n$ with the probabilities $p_1, p_2, \dots, p_n$. It is clear that $p_1 + p_2 + \cdots + p_n = 1$, since as a result of an experiment the variable X will have to take on one of the possible values. For example, if X is the face that showed on a die, the possible values of X will be $x_1 = 1$, $x_2 = 2, \dots, x_6 = 6$, while all the probabilities p_i $(i = 1, \dots, 6)$ are equal to $\frac{1}{6}$. Let us draw the following table.

x_i	1	2	3	4	5	6
p_i	$\frac{1}{6}$	$\frac{1}{6}$	$\frac{1}{6}$	$\frac{1}{6}$	$\frac{1}{6}$	$\frac{1}{6}$

Such a table, establishing a relationship between the possible values of a random variable and the respective probabilities, is known as the *law of distribution of a random variable* X; the law of distribution of a random variable is a functional relationship between the values of a *random variable* and the corresponding probabilities. Such a function fully describes a random variable in probabilistic terms. However, very often it is enough to find some numerical parameters describing the essential features of distribution of a random variable. Such numerical parameters are called its *numerical characteristics.*

We shall introduce some numerical characteristics of a random variable that are necessary for the further discussion of sport problems.

Mathematical expectation (*mean value*) of the discrete random variable X is the sum of the products of all of its possible values $x_1, \dots, x_n$ by the respective probabilities of these values $p_1, \dots, p_n$, i.e.,

$$M[X] = m_x = x_1 p_1 + x_2 p_2 + \cdots + x_n p_n,$$

where $M[X]$ and m_x denote the expectation of X.

It can be computed that in the example with the die $M[X] = 3.5$. Obviously, no such number can ever face up, yet this is the average possible number turning up.

The *variance* of a discrete random variable is the expectation of the square of the deviation of this variable from its expectation. Variance determines the extent of the scatter (deviation) of the possible values of a random variable relative to its expectation. Variance is denoted by $D[X]$ or D_x and is computed, by definition, from the formula

$$\begin{aligned} D[X] = D_x &= M[(m_x - X)^2)] \\ &= (m_x - x_1)^2 p_1 + (m_x - x_1)^2 p_2 + \cdots + (m_x - x_n)^2 p_n. \end{aligned}$$

In the example with the die $D_x = 8.75/3 \approx 2.92$.

Another numerical characteristic that is used very often is the *standard deviation* $\sigma_x = \sqrt{D_x}$. In the example with the die $\sigma_x \approx 1.71$.

Unfortunately, the probabilities of the possible values of a random variable are not usually known prior to experiments. The example with the die is an agreeable exception. On the other hand, for processing statistical data one has to know the numerical characteristics—such as expectation, deviation, etc.—of the random variable. Then what should be done if there is not any further information (e.g., based on symmetry) that would make it possible to infer the probabilities p_i? In such cases the answer can only be found from a series of trials. Let us explain the main point.

Suppose we have staged N independent trials in which each of the possible values x_i of the random variable X occurred m_i $(i = 1, \ldots, n)$ times. Let us make up the arithmetic mean m_x^* (sample mean) of the values of X observed in the trials:

$$m_x^* = \frac{x_1 m_1 + m_2 x_2 + \cdots + m_n x_n}{m_1 + m_2 + \cdots + m_n} = \sum_{i=1}^{n} x_i \frac{m_i}{N} = \sum_{i=1}^{n} x_i p_i^*.$$

The quantities $p_i^* = m_i/N$ $(i = 1, \ldots, n)$ are known as the frequency (*statistical probability*) of occurrence of the value x_i of the variable X in N trials. We have already noted (see 3.5) that, with the growing number N of trials, the frequency p_i^* differs less and less from the probability p_i. Therefore, the arithmetic mean m_x^* of the observed values of X will also differ less and less from the expectation m_x; in other words, m_x^* converges in probability to m_x. The quantity m_x^* is called *statistical mathematical expectation* or mean.

The relationship established between the arithmetic mean and expectation also is one of the manifestations of the Bernoulli law of large numbers, to which we already referred.

With many problems, an important part belongs to the *centered random variable* $\overset{\circ}{X} = X - m_x$, setting the deviation of X from its expectation m_x.

If we represent the values $x_1, \ldots, x_n$ of the random variable X by corresponding points on the numerical axis, then the centering of X (i.e., transition from X to $\overset{\circ}{X}$) signifies the transfer of the origin of ordinates to the point with the x-coordinate m_x.

It is obvious that the expectation of a centered random variable is equal to zero:

$$\begin{aligned} M[\overset{\circ}{X}] &= M[X - m_x] = \sum_{i=1}^{n} (x_i - m_x) p_i \\ &= \sum_{i=1}^{n} x_i p_i - m_x \sum_{i=1}^{n} p_i = m_x - m_x \cdot 1 = 0. \end{aligned}$$

Having recalled the definition, let us note that the variance D_x of the random variable X is the expectation of the square of the centered random variable $\overset{\circ}{X}$:

$$D_x = M[(X - m_x)^2] = M[\overset{\circ}{X}{}^2].$$

If in this expression we substitute for the expectation m_x the statistical mean m_x^*, we shall have the quantity

$$D_x^* = \sum_{i=1}^{n}(x_i - m_x^*)^2 p_i^*,$$

known as the *statistical variance of the random variable* X.

By processing the results of the experiment, we obtain the statistical values m_x^* and D_x^* of expectation and variance.

5.2. Faster! Let us discuss the problem of predicting an athlete's results at a competition given those achieved during preparation. Since any predicted value of a parameter, computed from a limited number of trials, will always comprise an element of chance, it is called the *estimator of a corresponding parameter*.

Thus, the estimator of the time t shown by a sprinter in a 100-meter dash will be denoted by $\tilde{t}$.

Let us first consider a more general problem. Let there be a random variable X whose distribution law contains an unknown parameter t (usually, expectation and variance). We have to find a suitable estimator $\tilde{t}$ of the value of this parameter from the results of n independent trials in which X takes on, respectively, the values $x_1, \ldots, x_n$. These values may be regarded as n "copies" of X. Each of the random variables X_i is distributed under the same law as X. It is quite obvious that the estimator $\tilde{t}$ is a function of $x_1, \ldots, x_n$ and, thus, is itself a random variable.

To be "good", the estimator $\tilde{t}$ of the parameter t must satisfy certain requirements. First of all, it is natural to require that, as the number of trials increases, the estimator $\tilde{t}$ should come closer (converge in probability) to the value of the parameter t. The estimator that has this quality is called *consistent*. Secondly it is desirable that, while we use the estimator $\tilde{t}$ instead of t, we should avoid systematic error, i.e., that the equality $M[\tilde{t}] = t$ be satisfied for the expectation of $\tilde{t}$. The estimator that satisfies this condition is called *unbiased*. Finally, we shall require the selected estimator $\tilde{t}$ to have the least variance of all; it is called an *efficient* estimator. It is not always possible to have all these requirements satisfied simultaneously. For example, though there may be an efficient estimator, the formulas for computing it may be too involved, compelling the use of an estimator with a somewhat greater variance.

The natural choice for the estimator $\tilde{m}_x$ of the expectation m_x is the arithmetic mean of the experimentally found values of the random variable X, i.e., $\tilde{m}_x = m_x^* = \frac{1}{n}(x_1 + \cdots + x_n)$.

As one learns from probability theory, this estimator is consistent, unbiased, and—given certain assumptions on the mode of the distribution law of the random variable X—efficient as well. As an estimator $\widetilde{D}_x$ of the variance

D_x, we choose the quantity

$$\widetilde{D}_x = \frac{n}{n-1}D_x^* = \frac{\sum_{i=1}^{n}(x_i - \tilde{m}_x)^2}{n-1}$$
$$= \frac{n}{n-1}\left[\frac{x_1^2 + x_2^2 + \cdots + x_n^2}{n} - \tilde{m}^2\right],$$

which is its consistent and unbiased estimator. The estimator $\tilde{t}$ of the parameter t (specifically, the estimators $\tilde{m}$ and $\widetilde{D}$ of expectation and variance) is expressed by a single number and is called a *point* estimator. If the volume of statistical data is not substantial, i.e., if the number n of outcomes of independent trials is small, the point estimator may differ a great deal from the parameter being estimated, often resulting in significant error. Therefore, when n is small, an *interval estimator* is used. It is defined by two numbers—the ends of the estimation interval.

Referring the reader for details to books on probability theory, we shall present here merely some general arguments leading to the concept of interval estimator. Assume that for the parameter t an unbiased estimator $\tilde{t}$ has already been found. it is clear that the estimator approximates the parameter more accurately according to a less absolute value of the difference $|t - \tilde{t}|$. Once for a positive number ε the inequality $|t - \tilde{t}| < \varepsilon$ is satisfied, it is natural to take ε as the measure of the estimator's accuracy. It should be borne in mind, however, that in mathematical statistics one cannot assert that for the estimator ε the inequality $|t - \tilde{t}| < \varepsilon$ is sure to be satisfied. It makes sense merely to speak of the probability $P(|t - \tilde{t}| < \varepsilon)$ that this inequality will be satisfied.

The probability $\alpha = P(|t - \tilde{t}| < \varepsilon)$ that the inequality $|t - \tilde{t}| < \varepsilon$ is satisfied is called the *fiducial probability* (*reliability*) of estimating t with the help of $\tilde{t}$.

Usually fiducial probability is set (assigned) in advance. For example, α is assumed to be equal to 0.85, 0.90, 0.999, or some other number close to one.

So, let us stipulate that, for some yet-unknown value of ε_α, the requirement $P(|t - \tilde{t}| < \varepsilon_\alpha) = \alpha$ is satisfied. This requirement is equivalent to $P(\tilde{t} - \varepsilon_\alpha < t < \tilde{t} + \varepsilon_\alpha) = \alpha$. The latter should be understood in the sense that the probability that the point t will get into the interval $I_\alpha = (\tilde{t} - \varepsilon_\alpha; \tilde{t} + \varepsilon_\alpha)$ is equal to α. Note that I_α itself is a random interval, since its middle $\tilde{t}$ on the number axis is in a random position, and it has a random length, $2\varepsilon_\alpha$, calculated from experimental data.

The interval I_α is called a *confidence interval*, and its ends are called *confidence bounds*. A confidence interval can naturally be considered as the range of possible values of the parameter t, consistent with the trial data (not at odds with them). Simultaneously, the probability that the inequality $|t - \tilde{t}| > \varepsilon_\alpha$ is satisfied—i.e., the probability that the point t does not fall within the interval I_α—is $1 - \alpha$, and the closer α is to 1, the closer this probability is to zero. Since here we cannot go into the justification of the method of defining ε_α when α is assigned, we shall just "prescribe" it.

Let us assume, as before, that n independent trials are carried out with the random variable X and that its values, $x_1, \ldots, x_n$, are found. We find for the unknown numerical characteristics of X—expectation and variance—the respective estimators

$$\tilde{m}_x = \frac{1}{n}\sum_{i=1}^{n} x_i, \qquad \widetilde{D}_x = \frac{1}{n-1}\sum_{i=1}^{n}(x_i - \tilde{m}_x)^2.$$

It can be proved that for the assigned probability α, the value of ε_α can be found by the formula

$$\varepsilon_\alpha = \sqrt{\frac{\widetilde{D}_x}{n}}\arg\Phi^*\left(\frac{1+\alpha}{2}\right),$$

where $\arg\Phi^*((1+\alpha)/2)$ is a function inverse to the Laplace function. In other words, $\arg\Phi^*((1+\alpha)/2)$ is the value of the argument at which the value of the Laplace function is equal to $(1+\alpha)/2$.

To make it easier to perform further calculations (including those the reader may like to do on his own), we shall cite here a short table of values of the function $\arg\Phi^*((1+\alpha)/2)$.

TABLE 1

α	ε_α	α	ε_α	α	ε_α	α	ε_α
0.80	1.282	0.86	1.475	0.92	1.750	0.97	2.169
0.81	1.310	0.87	1.513	0.93	1.810	0.98	2.325
0.82	1.340	0.88	1.554	0.94	1.880	0.99	2.576
0.83	1.371	0.89	1.597	0.95	1.960	0.9973	3.000
0.84	1.404	0.90	1.643	0.96	2.053	0.999	3.290
0.85	1.439	0.91	1.694				

Assume that our runner, in the last month before the competition, made twenty 100-meter dashes and showed the following results: 10.5, 10.8, 11.2, 10.9, 10.4, 10.6, 10.9, 11.0, 10.3, 10.8, 10.6, 11.3, 10.5, 10.7, 10.8, 10.9, 10.8, 10.7, 10.9, 11.0. We have to estimate the expected competition result and find the confidence interval for the fiducial probability $\alpha = 0.8$.

First we find the statistical values of $\tilde{m} = 10.78$ and $\widetilde{D} = 0.064$. Then we find from the table $\arg\Phi^*((1+0.8)/2) = 1.282$. Thus, $\varepsilon_\alpha = \sqrt{0.064/20} \cdot 1.287 = 0.072$.

Therefore, the confidence interval bounds are equal to $\tilde{m} - 0.072 = 10.71$ and $\tilde{m} + 0.072 = 10.85$.

Thus, with probability 0.8 the runner's result will be between 10.71 and 10.85. If we take as fiducial probability $\alpha = 0.9$, we can find that $\varepsilon_\alpha = 0.93$, and so the runner's expected result will be between 10.52 and 11.06.

5.3. Higher! It would be interesting to know what results pole vaulters will show in 1990. Mathematical statistics can help us answer (of course, in a first approximation) this as well. First, however, we must introduce the concepts of a system of random variables and covariance.

In mathematical statistics, problems often have more than one variable describing the outcome of a trial. This is the case of a *system of random variables.* One example of it is a system of two variables, H and Q, being a person's height and weight, respectively. The pair H, Q is a system of random variables, since an individual's exact weight and height are not known in advance (before measurement).

Now consider a pair of random variables X, Y. As before, let X take on the values $x_1, \dots, x_n$ with probabilities $p_1, \dots, p_n$, and let the random variable Y take on the values $y_1, \dots, y_l$, with probabilities $q_1, \dots, q_l$. The corresponding centered random variables will be denoted by $\overset{\circ}{X}$ and $\overset{\circ}{Y}$. Let us introduce on the plane a Cartesian system of coordinates, xOy, and consider a point C_0 with coordinates $x_0 = m_x$ and $y_0 = m_y$. The values (x_i, y_j) of the random variables (X, Y) are represented by points scattered around C_0.

An important numerical characteristic of a pair of random variables (characterizing, as it were, the aforementioned scatter) is covariance K_{xy} of the variables X, Y. The *covariance* K_{xy} of the random variables X, Y is the expectation of the product $\overset{\circ}{X}\overset{\circ}{Y}$ of centered random variables corresponding to X and Y:

$$K_{xy} = M[\overset{\circ}{X}\overset{\circ}{Y}] = M[(X - m_x)(Y - m_y)].$$

The covariance K_{xy} is a characteristic of the relationship between X and Y. Let us transform the expression for K_{xy}. To this end, we shall denote by p_{ij} the probability of joint occurrence of the values $X = x_i$, $Y = y_j$ of the random variables X and Y. All possible values of the product $\overset{\circ}{X}\overset{\circ}{Y}$ are expressed by the variables $(x_i - m_x)(y_j - m_y)$, and the expectation, i.e., covariance, will be expressed by the sum of products

$$K_{xy} = \sum_{j=1}^{l}\sum_{i=1}^{n}(x_i - m_x)(y_j - m_y)p_{ij}.$$

Assume, in particular, that X and Y are independent variables (see 3.5). Then the probability p_{ij} of the occurrence of the pair (x_i, y_j) is equal to the product $p_i q_j$ of the probabilities of the independent occurrence of the values $X = x_i$, $Y = y_j$. In this case, changing the order of summation, we shall find

$$\begin{aligned} K_{xy} &= \sum_{j=1}^{l}\sum_{i=1}^{n}(x_i - m_x)(y_j - m_y)p_i q_j \\ &= \sum_{i=1}^{n}(x_i - m_x)p_i \sum_{j=1}^{l}(y_j - m_y)q_j = M[\overset{\circ}{X}]M[\overset{\circ}{Y}] = 0, \end{aligned}$$

since the expectation of each centered random variable is equal to zero. Hence, for independent random variables, covariance is equal to zero.

Besides, from the formula for K_{xy} we can infer that, once the values of one of the random variables, say X, deviate insignificantly from their expectation, then the covariance also will be small, no matter how close the dependence between X and Y.

Instead of K_{xy}, one usually considers the *correlation coefficient*

$$r_{xy} = K_{xy}/\sigma_x\sigma_y,$$

where σ_x, σ_y are the standard deviations of X and Y.

If the covariance K_{xy} (or, which is the same thing, the correlation coefficient r_{xy}) equals zero, the variables X and Y are *uncorrelated*.

Independent random variables are always uncorrelated. However, the converse is generally not true: if r_{xy} is equal to zero, it does not follow that X and Y are independent.

The dependence between X and Y may be highly involved and not explicit. We shall dwell on the simplest yet important case, where it is known in advance that X and Y are linearly dependent (a and b are constants): $Y = aX + b$. Let us compute for this dependence the correlation coefficient:

$$\begin{aligned} K_{xy} &= M[\overset{\circ}{X}\overset{\circ}{Y}] = M[(X - m_x)(Y - m_y)] \\ &= M[(X - m_x)(aX + b - am_x - b)] = M[(X - m_x)^2 a] \\ &= aM[(X - m_x)^2] = aD[X]. \end{aligned}$$

Let us, further, find the covariance of Y:

$$D[Y] = D[aX + b] = D[aX] = a^2 D[X].$$

From this we obtain, for the standard deviations $\sigma_x = \sqrt{D_x}$, $\sigma_y = \sqrt{D_y}$ a similar dependence; $\sigma_y = |a|\sigma_x$.

Hence, the correlation coefficient

$$r_{xy} = \frac{K_{xy}}{\sigma_x\sigma_y} = \frac{aD_x}{\sigma_x|a|\sigma_x} = \frac{a\sigma_x^2}{|a|\sigma_x^2} = \frac{a}{|a|} = \begin{cases} 1 & \text{for } a > 0, \\ -1 & \text{for } a < 0. \end{cases}$$

Thus, if the dependence between X and Y is linear, then $r_{xy} = \pm 1$, according to the sign of the coefficient a.

It can be proved that in case of arbitrary random dependence between X and Y, the correlation coefficient is subject to inequalities $-1 \le r_{xy} \le 1$.

Indeed, let us introduce a random variable $U = \sigma_y X \pm \sigma_x Y$—a linear combination of X and Y with the coefficients σ_y, $\pm\sigma_x$. We do not know the mode of random dependence between X and Y. Let us compute the variance $D[U]$. To this end, let us satisfy ourselves that the variance of the sum $X' + Y'$ of two random variables is equal to the sum of their variances added to the doubled value of the covariance:

$$D[X' + Y'] = D[X'] + D[Y'] + 2K_{xy}.$$

To verify it, let us introduce the notation $Z = X' + Y'$. Since the expectation of a sum of random variables is the sum of their expectations— $m_z = m_{x'} + m_{y'}$—so $Z - m_z = X' - m_{x'} + Y' - m_{y'}$, i.e., for centered variables the same relationship occurs: $\overset{\circ}{Z} = \overset{\circ}{X} + \overset{\circ}{Y}$. Hence,

$$D[X' + Y'] = M[\overset{\circ}{Z}{}^2] = M[\overset{\circ}{X}{}'^2 + \overset{\circ}{Y}{}'^2 + 2\overset{\circ}{X}{}'\overset{\circ}{Y}{}']$$
$$= M[\overset{\circ}{X}{}'^2] + M[\overset{\circ}{Y}{}'^2] + 2M[\overset{\circ}{X}{}'\overset{\circ}{Y}{}'] = D_x + D_y + 2K_{xy}.$$

Thus, we have obtained the required result.

It now remains to substitute $X' = \sigma_y X$, $Y' = \sigma_x Y$ and note that

$$D[\sigma_y X \pm \sigma_x Y] = D[\sigma_y X] + D[\sigma_x Y] \pm 2M[\sigma_x \sigma_y \overset{\circ}{X}\overset{\circ}{Y}]$$
$$= \sigma_y^2 D_x + \sigma_x^2 D_y \pm 2\sigma_x \sigma_y K_{xy}.$$

But the variance of an arbitrary random variable is nonnegative:

$$\sigma_y^2 D_x + \sigma_x^2 D_y \pm 2\sigma_x \sigma_y K_{xy} \geq 0.$$

Since $D_x = \sigma_x^2$, $D_y = \sigma_y^2$,

$$2\sigma_x^2 \sigma_y^2 \pm 2\sigma_x \sigma_y K_{xy} \geq 0, \qquad \sigma_x \sigma_y \pm K_{xy} \geq 0,$$

or

$$1 \pm \frac{K_{xy}}{\sigma_x \sigma_y} \geq 0,$$

i.e., $|r_{xy}| \leq 1$, or $-1 \leq r_{xy} \leq 1$.

Let us represent in the system of coordinates xOy the observed pairs of values (x_i, y_j) as corresponding points. From the arrangement of these points we can judge, in a first approximation, if there is any correlation (specifically, linear correlation) between X and Y.

In the general case, as we saw, the correlation coefficient varies within the bounds: $-1 \leq r_{xy} \leq 1$. If $r_{xy} > 0$, the correlation is positive; if $r_{xy} < 0$, it is negative. The fact of positive correlation between random variables implies that if one of them increases, the other tends, on the average, to increase too; and in the case of negative correlation, if one of the variables increases, the other tends, on the average, to decrease. If we take, for example, a person's height and weight, we naturally deal with a positive correlation. It is also natural that the greater the coefficient's absolute value, the closer the relationship between the appropriate random variables.

Now we are ready to return to the question posed at the beginning of this subsection. To this end, we shall try to construct a linear relationship between records and the years they were achieved and to find the correlation coefficient, thus establishing the accuracy of the prediction. Let us write a table of world records in pole vaulting **[12]**:

4.78	R. Gutowski (USA)	1957
4.80	D. Bragg (USA)	1960
4.83	G. Davies (USA)	1961
4.89	J. Uelses (USA)	1962
4.93	D. Tork (USA)	1962
4.94	P. Nikula (Finland)	1962
5.00	B. Sternberg (USA)	1963
5.08	B. Sternberg	1963
5.13	J. Pennel (USA)	1963
5.20	J. Pennel	1963
5.23	F. Hansen (USA)	1964
5.28	F. Hansen	1964
5.32	R. Seagren (USA)	1966
5.34	J. Pennel	1966
5.36	R. Seagren	1967
5.38	P. Wilson (USA)	1967
5.41	R. Seagren	1968
5.44	J. Pennel	1969
5.45	W. Nordwig (DDR)	1970
5.46	W. Nordwig	1970
5.49	C. Papanicolaou (Greece)	1970
5.51	K. Isaksson (Sweden)	1972
5.54	K. Isaksson	1972
5.55	K. Isaksson	1972
5.63	R. Seagren	1972
5.65	D. Roberts (USA)	1975
5.67	E. Bell (USA)	1976
5.70	D. Roberts	1976
5.72	W. Kozakiewicz (Poland)	1980
5.75	T. Vigneron (France)	1980
5.75	T. Vigneron	1980
5.77	Ph. Houvion (France)	1980
5.78	W. Kozakiewicz	1980
5.80	T. Vigneron	1981
5.81	V. Poljakov (USSR)	1981
5.82	P. Quinon (France)	1983
5.83	T. Vigneron	1983
5.85	S. Bubka (USSR)	1984
5.88	S. Bubka	1984
5.90	S. Bubka	1984
5.94	S. Bubka	1984
6.00	S. Bubka	1985

6.01 S. Bubka 1986
6.03 S. Bubka 1987

The data from this table can be represented graphically (see Figure 11).

To construct a linear relationship between the record and the year, we shall use the least squares method. Roughly, it is a method of finding a line such that the sum of squared deviations of the observed points from the line is the least. In fact, we want to construct a linear function that smooths out, as it were, the observed (actually nonlinear) relationship between the random variables X and Y.

Without going into the technical details of calculation of the coefficients a and b, which define such a line $Y = aX + b$, we shall just state that $a = \widetilde{K}_{xy}/\widetilde{D}_x$, $b = \tilde{m}_y - a\tilde{m}_x$, where $\tilde{m}_x = \frac{1}{n}\sum_{i=1}^n x_i$, $\tilde{m}_y = \frac{1}{n}\sum_{i=1}^n y_i$, $\widetilde{D}_x$ and $\widetilde{D}_y$ are variances of X and Y.

In our example, $\tilde{m}_x \approx (19)73$ (we drop the first two digits in the designation of the year to simplify the computation process), $\widetilde{D}_x = 77.30$; the mean value of the world-record height attained is (in the time frame under consideration $\tilde{m}_y = (5)49\,\mathrm{m}$ (the figure for meters will again be omitted.) In addition, $\widetilde{D}_x = 1301$, $\widetilde{K}_{xy} = 306.26$, $\tilde{\sigma}_y = 36.07$, $\tilde{\sigma}_x = 8.79$, $r_{xy} = \widetilde{K}_{xy}/\tilde{\sigma}_x\tilde{\sigma}_y = 0.966$. With these values we consequently obtain $a \approx 3.96$ and $b \approx -240$, which gives the linear equation $y = 3.96x - 240$.

The correlation coefficient of 0.966 so obtained points to an almost linear dependence between X and Y. Accordingly the prediction based on a linear estimate can be regarded as relatively reliable, which—at least over a modest time period—permits us to expect a good estimated value.

In the case of application of the above data, the computation for the year 1990 yields $y = 116\,(\mathrm{cm})$, i.e. the expected world-record position for 1990 is $5\,\mathrm{m} + 116\,\mathrm{cm} = 6.16\,\mathrm{m}$. For $x = 95$ (i.e., for 1995) we find the value $y = 136$ from the above equation, so that the result conjectured for 1995 is consequently about 6.36 m. (One presumably obtains more realistic estimates if one includes in the investigation only the world record of the last 20 or 30

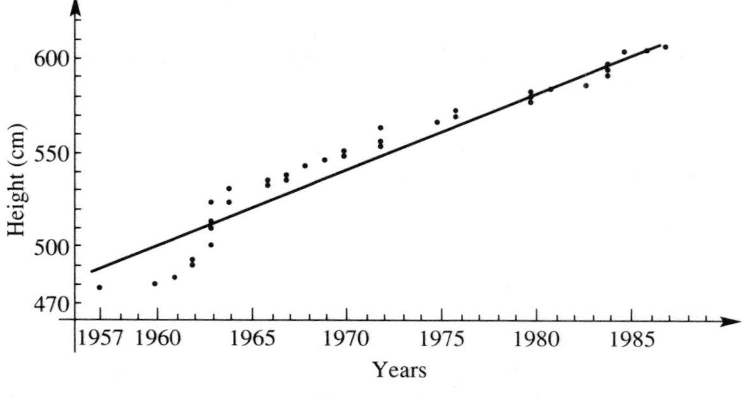

FIGURE 11

years, which will be left to the reader.) The future will show it! [18]

5.4. Stronger! In this section we shall go on with our introduction to correlation analysis, applying it to a sport problem. The purpose of this part of mathematical statistics is to investigate relationships between things or, rather, to find out whether there are such relationships. Here is the problem.

Let there be a system of random variables $X_1, X_2, \dots, X_m$. This system is subjected to n independent observations, and the results are tabulated (see Table 2). Each row of the table contains m values assumed by the random variables X_k $(k = 1, \dots, m)$ in a single observation.

TABLE 2

Trial Number	X_1	X_2	$\cdots$	X_k	$\cdots$	X_m
1	x_{11}	x_{12}	$\cdots$	x_{1k}	$\cdots$	x_{1m}
2	x_{21}	x_{22}	$\cdots$	x_{2k}	$\cdots$	x_{2m}
$\cdots$	$\cdots$	$\cdots$	$\cdots$	$\cdots$	$\cdots$	$\cdots$
i	x_{i1}	x_{i2}	$\cdots$	x_{ik}	$\cdots$	x_{im}
$\cdots$	$\cdots$	$\cdots$	$\cdots$	$\cdots$	$\cdots$	$\cdots$
n	x_{n1}	x_{n2}	$\cdots$	x_{nk}	$\cdots$	x_{nm}

The numbers in this table, with two indices each, are record observation results. The first index shows the number of the observation, and the second is the number of the random variables. Thus, x_{ik} is the value assumed by the random variable X_k on the ith trial.

We are going to find estimators for the numerical characteristics of this system and, above all, to find the elements of the covariance matrix

$$K = \begin{pmatrix} k_{11} & k_{12} & \cdots & k_{1m} \\ & k_{22} & \cdots & k_{2m} \\ \cdots & \cdots & \cdots & \cdots \\ & & \cdots & k_{mm} \end{pmatrix},$$

in which the $k_{ij} = k_{ji}$ are the covariances of the ith and jth random variables. By virtue of the symmetry of the matrix K relative to the principal diagonal, only the elements of the latter and those above it are written in the matrix, the other—the symmetrical part—being omitted from the notation.

Of most interest is the normalized covariance matrix comprised of correlation coefficients. From this matrix we can also gauge the extent of the relationship between the random variables and their impact upon one another. We shall denote this matrix by

$$r = \begin{pmatrix} 1 & r_{12} & & \cdots & r_{1m} \\ & 1 & r_{23} & \cdots & r_{2m} \\ \cdots & \cdots & \cdots & \cdots & \cdots \\ & & & \cdots & 1 \end{pmatrix},$$

[18] On September 6, 1988, at a sports festival in Bratislava, Sergei Bubka (USSR) improved the (then old) world record by two centimeters to 6.05 meters, and, subsequently, in Nice, on October 7, 1988, he added one more centimeter to the record: 6.06 meters. This is where the world record stood in Summer, 1990.

where $r_{ii} = 1$ is the correlation coefficient of X_i with itself, and

$$r_{ij} = k_{ij}/\sigma_i\sigma_j .$$

To determine the correlation coefficients from statistical data, let us first find estimators for the expectations

$$\tilde{m}_i = \frac{1}{n}\sum_{i=1}^{n} x_u \qquad (l = 1, \dots, m).$$

Next, we compute the unbiased estimators for the variances

$$\widetilde{D}_l = \frac{1}{n-1}\sum_{i=1}^{n}(x_u - \tilde{m}_l)^2 \qquad (l = 1, \dots, m)$$

and covariances

$$\tilde{k}_{ij} = \frac{1}{n-1}\sum_{i=1}^{n}(x_{li} - \tilde{m}_i)(x_{ji} - \tilde{m}_j) \quad (l = 1, \dots, m;\ j = 1, \dots, m).$$

From these data we define estimators for the elements of the normalized correlation matrix

$$\tilde{r}_{ij} = \tilde{k}_{ij}/\tilde{\sigma}_i\tilde{\sigma}_j \quad (l = 1, \dots, m;\ j = 1, \dots, m).$$

Let us now turn to weight lifting and employ the statistical techniques described above to show the dependence between the following quantities:

X_1 the sum of snatch, and clean and jerk (kg),
X_2 snatch (kg),
X_3 clean and jerk (kg)
X_4 athlete's weight (kg),
X_5 athlete's age (year of birth).

For our statistical input data, let us consider information about the world's top ten athletes and records; in this way we shall, as it were, stage ten trials. The data are drawn entirely from materials prepared by the international judge M. Aptekar and published in the daily "Sovetskiĭ Sport" ("Soviet Sports") on October 19, 1983.

Let us tabulate these materials (see Table 3).

TABLE 3

Athletes	X_1	X_2	X_3	X_4	X_5
1	257.5	115.5	140	52	1964
2	287.5	128	160.5	56	1961
3	305	137.5	173	60	1962
4	345	154	196	67.5	1959
5	367.5	165	209	75	1962
6	400	180	223.5	82.5	1959
7	420	195.5	228.5	90	1956
8	440	200	240.5	100	1962
9	435	196.5	243	110	1957
10	460	205.5	261	130	1961

We compute by the already familiar formulas

$$\begin{array}{lll}
\tilde{m}_1 = 371.75 \approx 372; & \widetilde{D}_1 = 4990; & \tilde{\sigma}_1 = 70.6; \\
\tilde{m}_2 = 167.75 \approx 168; & \widetilde{D}_2 = 1070; & \tilde{\sigma}_2 = 32.7; \\
\tilde{m}_3 = 208; & \widetilde{D}_3 = 1483; & \tilde{\sigma}_3 = 38.5; \\
\tilde{m}_4 = 82.3 \approx 82; & \widetilde{D}_4 = 644; & \tilde{\sigma}_4 = 25.4; \\
\tilde{m}_5 = 1960.3 \approx 1960; & \widetilde{D}_5 = 5.66; & \tilde{\sigma}_5 = 2.4.
\end{array}$$

Let us develop a table of differences $x_{i_l} - \tilde{m}_l$ $(l = 1, \ldots, 5)$ (Table 4) and their squares (Table 5).

TABLE 4

Athletes	$x_1 - \tilde{m}_1$	$x_2 - \tilde{m}_2$	$x_3 - \tilde{m}_3$	$x_4 - \tilde{m}_4$	$x_5 - \tilde{m}_5$
1	−114.5	−52.5	−63	−30	3
2	−84.5	−40	−47.5	−26	1
3	−67	−30.5	−35	−22	2
4	−27	−14	−12	−14.5	−1
5	−4.5	−3	1	−7	2
6	28	12	15.5	0.5	−1
7	48	27.5	20.5	8	−4
8	68	32	32.5	18	2
9	63	28.5	35	28	−3
10	88	37.5	53	48	1

TABLE 5

Athletes	$(x_1 - \tilde{m}_1)^2$	$(x_2 - \tilde{m}_2)^2$	$(x_3 - \tilde{m}_3)^2$	$(x_4 - \tilde{m}_4)^2$	$(x_5 - \tilde{m}_5)^2$
1	13110	2756	3969	900	9
2	7140	1600	2256	676	1
3	4489	930	1225	484	4
4	729	196	144	210	1
5	20	9	1	49	4
6	784	144	240	64	1
7	2304	756	420	324	16
8	4624	1024	1056	784	4
9	3969	812	1225	2304	9
10	7744	1406	2809		1

Let us find the covariances:

$$k_{11} = \widetilde{D}_1 = \frac{44913}{9} = 4990; \qquad k_{22} = \widetilde{D}_2 = \frac{9633}{9} = 1070;$$

$$k_{33} = \widetilde{D}_3 = \frac{13346}{9} = 1483; \qquad k_{44} = \widetilde{D}_4 = \frac{5796}{9} = 644;$$

$$k_{55} = \widetilde{D}_5 = \frac{5094}{9} = 566;$$

$$k_{12} = \frac{20753}{70.6 \cdot 32.7} = 0.90; \qquad k_{13} = \frac{22279}{70.6 \cdot 38.5} = 0.82;$$

$$k_{14} = 0.67; \quad k_{15} = -0.34; \quad k_{23} = 0.89; \quad k_{24} = 0.71;$$
$$k_{25} = -0.45; \quad k_{34} = 0.85; \quad k_{35} = 0.41; \quad k_{45} = -0.31.$$

This is what the normalized correlation matrix looks like in our case:

$$\tilde{r} = \begin{pmatrix} 1 & 0.9 & 0.82 & 0.67 & -0.34 \\ & 1 & 0.89 & 0.71 & -0.45 \\ & & 1 & 0.85 & -0.41 \\ & & & 1 & -0.31 \\ & & & & 1 \end{pmatrix}$$

It is easy to see that while X_1, X_2, X_3, X_4 are sufficiently strongly correlated, the year of birth is hardly correlated at all—there is merely a slight "aging" with the growth in weight category.

5.5. A jump into the twenty-first century? Bob Beamon (US) set a world record of 890 cm at the 1968 Olympic Games at Mexico City. This phenomenal achievement was called a jump into the twenty-first century.

Recently many athletes managed to jump to distances of 830–870 cm. Granted, such results are still rare: they are registered at most four times a year.

We shall try to estimate, if only approximately, the probability that Beamon's record will be exceeded even in this century.

We shall call "successful" any jump to a distance exceeding 830 cm. About thirty such jumps have been recorded in track and field history. The first successful jump (831 cm) was made by the Soviet athlete I. Ter-Ovanesian in 1962. Registered between 1962 and 1982 were fifteen successful jumps (besides Beamon's). Here is their list, compiled from the track-and-field reference book **[12]**:

831	I. Ter-Ovanesian (USSR)	1962
831	R. Boston (USA)	1964
834	R. Boston	1964
835	R. Boston	1965
835	I. Ter-Ovanesian	1967

835	J. Schwarz (FRG)	1970
835	A. Robinson (USA)	1976
835	H. Lauterback (GDR)	1981
836	J.-C. di Oliveira (Brazil)	1979
841	Sh. Abbyasov (USSR)	1982
841	L. Dombrowski (GDR)	1982
845	N. Stekić (Yugoslavia)	1975
852	L. Mirix (USA)	1979
854	L. Dombrowski	1980
876	C. Lewis (USA)	1982
890	R. Beamon	1968

We shall introduce a random variable L, equal to the length of a jump, rounded off to the nearest ten down. Let us divide the entire range of the marked values of L into 10-centimeter intervals (classes), compute the number m_i of values per each ith class, and then find the statistical probability $\tilde{p}_i = m_i/N$, dividing m_i by the total number $N = 15$ of the marked values. We arrive at a statistical series:

L	830	840	850	860	870	880	890	900
$\tilde{p}$	$\frac{8}{15}$	$\frac{4}{15}$	$\frac{2}{15}$	0	$\frac{1}{15}$	0	0	0

It can be presented graphically as a histogram (Figure 12).

It is easy to see that this series is consistent on different statistical criteria, which we shall not discuss here (see [**3**, **26**]), with the following theoretical distribution series of the random variable X:

X	830	840	850	860
p	$\frac{1}{2}$	$\frac{1}{4}$	$\frac{1}{8}$	$\frac{1}{16}$

870	880	890	900
$\frac{1}{32}$	$\frac{1}{64}$	$\frac{1}{128}$	$\frac{1}{256}$

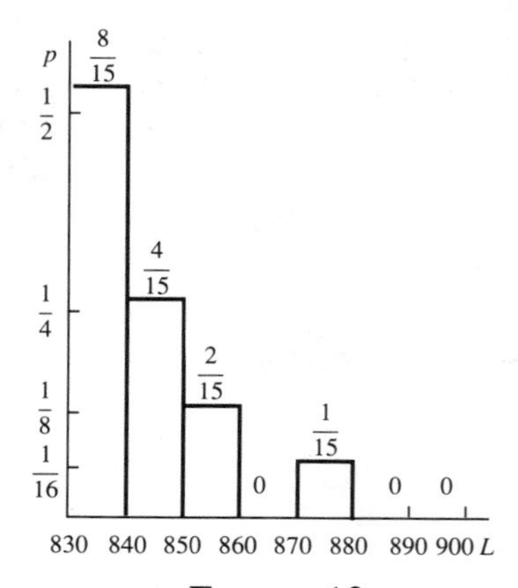

FIGURE 12

From this it follows that the probability of a "successful" jump, not exceeding Beamon's record, is equal to

$$p(X < 890) = \tfrac{1}{2} + \tfrac{1}{4} + \tfrac{1}{8} + \cdots + \tfrac{1}{64} = \tfrac{63}{64}.$$

If we assume that the greatest annual number of "successful" jumps is four, then according to this somewhat inflated estimate, we can expect sixty "successful" jumps before the century is out.

The probability q that none of them will top Beamon's record is equal to

$$q = (p(X < 890))^{60} = (\tfrac{63}{64})^{60} = 0.389.$$

Hence, the probability that at least one of these jumps is a new world record (i.e., exceeds 8.90 m) is $p = 1 - q = 0.611$. In fact, as follows from years of observation, only two successful jumps are registered on the average every year. Therefore, a more realistic assessment would be

$$q = (p(X < 890))^{30} = 0.625,$$

and the probability that Beamon's record will be exceeded in this century is $p = 0.375$.

Thus the probability that Bob Beamon has "jumped into the twenty-first century" is sufficiently great. [19]

---

[19] At international competitions in Los Angeles, CA, the holder of four 1984 Olympic gold medals C. Lewis, on May 18, 1985, set out to beat Beamon's record. However, no new record was set. Only in a fourth attempt did Lewis take 8.77 m, with the fair wind exceeding the permissible level ("Soviet Sports", May 21, 1985).

Editor's note: Beamon's record was broken recently.

6

Linear Programming and Sports

6.1. Positioning players on a basketball team. An experienced coach, who knows his players well, usually copes successfully with the task of assigning to them their positions in the team. The problem of how to use the spares in different combinations turns out to be more difficult, if many of the players are of the approximate same strength. Then even an experienced coach may find a mathematical model helpful.

To start with, let us take a look at a simple and not entirely unusual situation. Shortly before an important event, some of a team's players and even the coach are replaced. The new coach is less experienced and does not know the players very well. He has to assign positions in such a way as to provide for the team's most effective performance.

We shall try to help the new coach by resorting to operations research. To this end, we shall put the problem, formulated on a verbal level, into a more exact shape and begin to construct its mathematical model. If you know nothing about the players, you have nothing to decide on and may as well proceed by guess. Therefore, even limited information is welcome. It is necessary to find out early enough what the players' strong points are. Usually the players are given a series of tests to estimate their ability to be a good center, forward, or guard, on the left side or on the right side. The players—we shall call them A, B, C, D, E—are given marks for their performance.

A seasoned coach might say, "Why do it when each player has his position anyway? there's no point in making a forward play a guard's position." To a certain extent that is so, but if there are many spares, it is very difficult to form a team for a match. This problem is resolved in the same way as the foregoing simplified problem.

This method can also be applied when the coach has to decide whether to move in two centers or two guards (instead of one).

Let us tabulate the test results (in points).

The higher the mark, the more reason to assign the respective position to the player. For example, B promises to be a good center and guard but a

TABLE 6

Player	Guard	Center	Forward	Left side	Right side
A	3	4	2	2	1
B	4	5	3	1	3
C	4	3	1	1	1
D	3	1	2	2	2
E	1	3	1	2	1

poor left side, while D, on the whole, plays every position rather well but is less than good as center.

Let us remember the meaning of the numbers in Table 6 and turn to the mark matrix Γ:

$$\Gamma = \begin{pmatrix} 3 & 4 & 2 & 2 & 1 \\ 4 & 5 & 3 & 1 & 3 \\ 4 & 3 & 1 & 1 & 1 \\ 3 & 1 & 2 & 2 & 2 \\ 1 & 3 & 1 & 2 & 1 \end{pmatrix}.$$

Along with the coach, we shall make the natural assumption (*effectiveness criterion*) that the team's effectiveness is determined by the total of the points evaluating the performance of each player. This assumption can be disputed and a different effectiveness criterion insisted on. The reader is welcome to do it and suggest a better alternative. It is almost certain to be more involved. The great merit of the criterion we have chosen is that it depends linearly on each player's marks. The meaning of this will be clear later on. In the meantime let us consider a specific suggestion (rash, if anything), namely, to position A as guard, B as center, C as forward, D as left side, and E as right side. With this arrangement P, the team's *effectiveness* (we shall denote it by $\Phi(P)$), expressed in points, will be

$$\Phi(P) = 3 + 5 + 1 + 2 + 1 = 12.$$

The arrangement P is corresponded to by Table (matrix) 7. It is an *assignment matrix*, corresponding to the arrangement P. We shall denote it also by P.

TABLE 7

Player	Guard	Center	Forward	Left Side	Right Side
A	1	0	0	0	0
B	0	1	0	0	0
C	0	0	1	0	0
D	0	0	0	1	0
E	0	0	0	0	1

The meaning of this table is clear: 1 at the intersection of A's row and the "Guard" column means that this position has been assigned to A; 0 confirms that the respective position has not been assigned to A. It is easy to see that each row and column of the assignment matrix has exactly one unit, while the rest of the elements are zeros. This structure reflects the stringent requirement that each player be assigned just one position and each position be assigned to just one player. These are as many possible assignment matrices—every possible arrangement of players in a team—as there are permutations of five, namely, $5! = 1\cdot2\cdot3\cdot4\cdot5 = 120$. Out of all these matrices, one has to choose a matrix P^* (there may be several of them) that will define the arrangement of players with the highest effectiveness value, compared with other assignment matrices P. We shall write down this requirement as $\Phi(P^*) = \max_P \Phi(P)$.

The number of all possible alternatives is only 120, and after looking them over, our coach finds the assignment matrix

$$P^* = \begin{pmatrix} 0 & 1 & 0 & 0 & 0 \\ 0 & 0 & 0 & 0 & 1 \\ 1 & 0 & 0 & 0 & 0 \\ 0 & 0 & 1 & 0 & 0 \\ 0 & 0 & 0 & 1 & 0 \end{pmatrix},$$

which promises the best results for the team, $\Phi(P^*) = 4+3+4+2+2 = 15$. Under this arrangement, A is center, B right side, C guard, D forward, and E left side.

This, however, is not the only optimal decision. The assistant coach discovers that the same maximum effectiveness value occurs when the players are positioned according to the assignment matrix

$$\widetilde{P} = \begin{pmatrix} 0 & 0 & 1 & 0 & 0 \\ 0 & 1 & 0 & 0 & 0 \\ 1 & 0 & 0 & 0 & 0 \\ 0 & 0 & 0 & 0 & 1 \\ 0 & 0 & 0 & 1 & 0 \end{pmatrix}; \qquad \Phi(\widetilde{P}) = 2+5+4+2+2 = 15.$$

Thus, the coach has solved the problem simply by exhaustion. He was able to do it because there were few alternatives (small dimension of the problem). The situation drastically changes if there are spares on the team, whose performance—just as that of the basic players—varies depending on who their partners are. This, too, could be taken into account, but it would make the model more involved, which is undesirable at this stage. So, we shall assume that the testing yields certain average marks, allowing for play with different partners. Even if there is one spare for each position on the team—i.e., if the total number of players is ten—the assignment problem requires an exhaustive search of, generally speaking, $10! = 3628800 \approx 3.6 \cdot 10^6$ alternatives. A direct exhaustive search in this case is impossible, unless it is done by means of a computer. Let us estimate the time. There are $3\cdot 10^7$ seconds in a year. At a rate of a search a second, it will take $1/10$ of a year; at a rate of a search a millisecond, it will take $1/10^4$ of a year; at a rate of a search a microsecond, it will take $1/10^7$ of a year; i.e., 3 seconds. But an exhaustive search of $20!$ alternatives (or more than $10^{10}\cdot 10!$) will take all of 3000 years. Not very encouraging, is it? Fortunately, the assignment problem (also known as the problem of choice) has a good mathematical model. This model is formalized in terms of linear programming—the most complete and widely applied part of mathematical programming or operations research theory.

Let us construct a mathematical model of the assignment problem. For the sake of convenience, we shall give the players A, B, C D, E numbers $i = 1, 2, 3, 4, 5$, respectively. Likewise, we shall number $j = 1, 2, 3, 4, 5$ the positions of guard, center, forward, left side, and right side, respectively. Then, we introduce 25 unknowns x_{ij} $(i = 1, \ldots, 5, j = 1, \ldots, 5)$, whose values we shall interpret as indications of the player i being assigned the position j. Furthermore, each of the variables x_{ij} can assume just one of two possible values: $x_{ij} = 1$, if the player i has been assigned the position j; $x_{ij} = 0$, otherwise.

The set of as yet unknown values of x_{ij} constitutes the assignment matrix

$$X = \begin{pmatrix} x_{11} & x_{12} & x_{13} & x_{14} & x_{15} \\ x_{21} & x_{22} & x_{23} & x_{24} & x_{25} \\ x_{31} & x_{32} & x_{33} & x_{34} & x_{35} \\ x_{41} & x_{42} & x_{43} & x_{44} & x_{45} \\ x_{51} & x_{52} & x_{53} & x_{54} & x_{55} \end{pmatrix}.$$

We have already encountered numerical realizations of such matrices in the foregoing example. We know that only one of the elements in each row and column of the matrix X is equal to 1, the rest being zeros. This compulsory condition (constraint) can be written as follows: the sum of the elements in each row (column) is equal to 1:

$$x_{11} + x_{12} + x_{13} + x_{14} + x_{15} = 1,$$
$$\ldots\ldots\ldots\ldots\ldots\ldots\ldots\ldots$$
$$x_{51} + x_{52} + x_{53} + x_{54} + x_{55} = 1,$$

$$x_{11} + x_{21} + x_{31} + x_{41} + x_{51} = 1,$$
$$\ldots\ldots\ldots\ldots\ldots\ldots\ldots\ldots$$
$$x_{15} + x_{25} + x_{35} + x_{45} + x_{55} = 1.$$

In a shorter notation,

$$\sum_{j=1}^{5} x_{ji} = 1 \qquad (i = 1, \ldots, 5), \tag{1}$$

$$\sum_{i=1}^{5} x_{ij} = 1 \qquad (j = 1, \ldots, 5). \tag{2}$$

We must add to this the requirement that the unknowns be nonnegative

$$x_{ij} \geq 0 \qquad (i = 1, \ldots, 5;\ j = 1, \ldots, 5). \tag{3}$$

The player i, assigned the position j, will contribute his share $a_i x_{ij}$ to the overall effectiveness $\Phi(X)$. Here a_{ij} is an element of the corresponding matrix of marks Γ, which lies at the intersection of the ith row and jth column of the matrix. The team's overall effectiveness is a total of 25 summands

$$\Phi(X) = \sum_{i=1}^{5} \sum_{j=1}^{5} a_{ij} x_{ij}. \tag{4}$$

In our concrete example (see Table 6)

$$\Phi(X) = 3x_{11} + 4x_{12} + \cdots + x_{55}. \tag{5}$$

The search for the assignment matrix X which gives the effectiveness $\Phi(X)$ its maximum value reduces to the following mathematical problem: Of all nonnegative solutions

$$x_{ij} \geq 0 \qquad (i = 1, \ldots, 5; \ j = 1, \ldots, 5)$$

of the constraint systems (1) and (2), choose that which gives function (4) the greatest value (optimizes $\Phi(X)$).

The problem thus formulated is the mathematical model of the assignment problem in a basketball team (without spares).

Assume that the team has $n > 5$ players. Then we introduce, in addition to the known five, another $k = n - 5$ dummy positions in the team, assuming that in each of them the test mark a_{ij} $(i = 1, \ldots, n; \ j = 6, 7, \ldots, n)$ of each player is equal to zero. Having done this, we arrive at the already familiar problem of choice, given an equal number of candidates and positions on the team. Now we have a mathematical model, differing from (1)–(4) only in the number variables x_{ij} and constraints.

Similarly, we can formulate and compute different versions of problems in which, for example, some positions are retained for the basic team and the rest are distributed among the spares.

The general assignment problem can be solved by the simplex method (see 6.8). It is more practical, however, to use the "Hungarian method", proposed by Egervari (1931), which successfully utilizes the problem's specific constraints. The reader will find this method described, for example, in **[17, 18]**.

6.2. Soccer clubs and players.[20] The Record sports firm owns three soccer clubs— B_1, B_2, and B_3.

As is fairly common in many countries, teams are partly replenished by purchasing players trained at junior sports centers and clubs of other organizations and countries. To simplify matters, let us assume that there are two

[20] Here is an excerpt from a review, "The Italian Legion", printed in the "Soviet Sports" on September 23, 1984: "Italy is perhaps the world's largest 'importer' of soccer 'stars'. The local clubs have bought many outstanding European and South American players. It's just as well, soccer columnists sarcastically observed, that Italian clubs are permitted to have not more than two foreign players. Of the sixteen Major League teams, 'Cremonese' is the only one without a single foreign player."

such centers, A_1 and A_2, and that in the current year these centers offer Record a_1 and a_2 players, respectively, of about the same strength. Simultaneously, the clubs need to replenish their memberships with b_1, b_2, and b_3 players, respectively. Without any loss of generality, we can assume that

$$a_1 + a_2 = b_1 + b_2 + b_3, \tag{6}$$

i.e., that the number of players trained at both centers coincides with the clubs' requirements. If this condition is violated—as will be clear later on—we can introduce a dummy club (when the output exceeds the demand) or a dummy center (when the demand exceeds the output).

Assume that center A_i hands over a player to club B_j for fixed sum of money c_{ij}. The problem for Record is to devise a plan for engaging players from the centers to play for the clubs in such a manner as to have all the newly trained players engaged and the clubs' requirements completely satisfied at the least expense.

Let us express the problem in mathematical terms. Denote by x_{ij} the yet-unknown number of players transferring from center A_i to club B_j. Then the total number of players transferring from centers A_1 and A_2 to club B_j is $x_{1j} + x_{2j}$. This total should coincide with the demand b_j of club B_j $(j = 1, 2, 3)$. Hence, we arrive at three equations:

$$x_{11} + x_{21} = b_1, \qquad x_{12} + x_{22} = b_2, \qquad x_{13} + x_{23} = b_3. \tag{7}$$

The total number of players turned over to the clubs by the centers is

$$x_{11} + x_{12} + x_{13} = a_1, \qquad x_{21} + x_{22} + x_{23} = a_2. \tag{8}$$

The players being distributed among the clubs according to the plan $X = \{x_{ij};\ i = 1, 2;\ j = 1, 2, 3\}$, Record's total expenditure amounts to

$$\begin{aligned} S(X) = c_{11}x_{11} + c_{12}x_{12} + c_{13}x_{13} + c_{21}x_{21} \\ + c_{22}x_{22} + c_{23}x_{23}. \end{aligned} \tag{9}$$

Thus, we arrive at the following mathematical problem (a linear programming problem): Among all nonnegative solutions x_{ij} of the constraint system (7)–(8), find the one where the form $S(X)$ attains the least value.

It is easy to see that our previous assignment problem is of the same form as the present problem; there all a_i were equal to 1 and all b_j were equal to 1.

As an introduction to the techniques of solving this type of problem, let us first consider a numerical example. Suppose that $a_1 = 2$, $a_2 = 3$, $b_1 = 1$, $b_2 = 3$, $b_3 = 1$, respectively. Assume that the expenditures are: $c_{11} = 4$, $c_{12} = 9$, $c_{13} = 3$, $c_{21} = 4$, $c_{22} = 8$, $c_{23} = 1$. Condition (6) is satisfied, and the constraints (7)–(8) and the form (9) to be minimized will be:

(10) $$x_{11}+x_{12}=1, \qquad x_{12}+x_{22}=3, \qquad x_{13}+x_{23}=1;$$

(11) $$x_{11}+x_{12}+x_{13}=2, \qquad x_{21}+x_{22}+x_{23}=3;$$

(12) $$x_{ij}\geq 0 \qquad (i=1,2;\ j=1,2,3);$$

(13) $$S(X)=4x_{11}+9x_{12}+3x_{13}+4x_{21}+8x_{22}+x_{23};$$

Let us repeat that if condition (6) is satisfied, then the system of constraints (10)–(11) is always consistent. In fact, if we leave out the minimization of the form $S(X)$, it is quite obvious that there are countless ways in which to provide reinforcements for the clubs. The system (10)–(11) of five algebraic equations in six unknowns (its rank $r = 4$) can be resolved with respect to four unknowns (called *basic*), which can be expressed through the two remaining (*slack*) unknowns. Choosing x_{13}, x_{21}, x_{22}, x_{23} as basic and x_{11} and x_{12} as slack variables, we find

(14) $$x_{13}=2-x_{11}-x_{12}, \qquad x_{23}=-1+x_{11}+x_{12},$$
$$x_{21}=1-x_{11}, \qquad x_{22}=3-x_{12}.$$

For the form $S(X)$ we obtain the expression

(15) $$S(X)=33-2x_{11}-x_{12}.$$

Lastly, let us write the nonnegativity conditions for all the variables

(I) $$2-x_{11}-x_{12}\geq 0,$$

(II) $$1-x_{11}\geq 0,$$

(III) $$3-x_{12}\geq 0,$$

(IV) $$-1+x_{11}+x_{12}\geq 0,$$

(V) $$x_{11}\geq 0,$$

(VI) $$x_{12}\geq 0.$$

Before passing to the geometric interpretation of the problem and its solution, we shall introduce some basic concepts of linear programming.

First, let us take up the inequality (I). Introduce in the plane a Cartesian coordinate system. Designate the axes x_{12} and x_{11}. Substitute in (I) the equality sign for the inequality sign:

(I′) $$2-x_{11}-x_{12}=0.$$

One learns in high school that any linear equation—(I′) in particular—determines a line in a given coordinate system. Further, consider the linear function (*linear form*) equal to the left-hand side of the equation (I′):

(16) $$F=2-x_{11}-x_{12}.$$

Choose in the plane an arbitrary point A_0 with coordinates (x_{11}^0, x_{12}^0), and substitute them in the expression (16) for the form F. The latter will assume a numerical value

$$F(A_0)=F_0=2-x_{11}^0-x_{12}^0.$$

This number is the value of F at the point A_0.

Let us now take an arbitrary number C and consider a point set in the plane (a locus of points in the plane) in which the form F is equal to the number C:

$$2 - x_{11} - x_{12} = C.$$

The latter equation determines a straight line in the plane. It is called the *line of equal values of the form* F. In particular, the line (I') is the line of zero values of F. Two straight lines—lines of equal values

$$x_{11} + x_{12} + C_1 - 2 = 0, \qquad x_{11} + x_{12} + C_2 - 2 = 0,$$

corresponding to different values C_1 and C_2, do not intersect: they have equal slopes (but different intercepts) and are parallel.

Thus, the whole plane is "fibered" into countless parallel lines—the lines of equal values of F. Transition from points on one line to those on another is accompanied by a change in the value of F.

All that has been said earlier about form (16) holds for any other linear form $\Phi = a_1x_{11} + a_2x_{12} + a_0$ with any real coefficients a_0, a_1, a_2. All of it is true for the linear forms—the left-hand of the inequalities (II)–(VI). Therefore, from now on we shall examine any form Φ, in which, to shorten the notation, we shall designate x_{11} and x_{12} simply as x_1 and x_2, respectively:

$$\Phi = a_0 + a_1x_1 + a_2x_2. \tag{17}$$

We shall continue with our example, after some general discussion, in the next section.

6.3. Some basic concepts and facts. Consider in the plane, where we introduce a Cartesian coordinate system, two points, M_1 and M_2, and their radius vectors $\mathbf{r}_1 = \overline{OM_1}$ and $\mathbf{r}_2 = \overline{OM_2}$ (see Figure 13). The difference between $\mathbf{r}_1$ and $\mathbf{r}_2$ is equal to $\mathbf{r}_2 - \mathbf{r}_1 = \overline{M_1M_2}$. Suppose that λ is any number satisfying the inequality $0 \le \lambda \le 1$. If we multiply the vector $\overline{M_1M_2}$ by the scalar λ, we get a vector $\lambda\overline{M_1M_2} = \mathbf{s}$, collinear to $\overline{M_1M_2}$, pointing in the same direction, but shorter than $\overline{M_1M_2}$. It is clear that, should the initial point of the vector $\mathbf{s}$ coincide with the point M_1, its end M will be inside the line segment M_1M_2. When $\lambda = 0$, M coincides with M_1; and when $\lambda = 1$, it coincides with M_2. For any $0 \le \lambda \le 1$, the radius vector $\mathbf{r}$ of the point M is equal to the sum $\mathbf{r} = \mathbf{r}_1 + \mathbf{s}_1 = \mathbf{r}_1 + \lambda(\mathbf{r}_2 - \mathbf{r}_1)$. Thus, the radius vector $\mathbf{r}$ of any point M on the line segment between M_1 and M_2 is found by the formula

$$\mathbf{r} = \mathbf{r}_1 + \lambda(\mathbf{r}_2 - \mathbf{r}_1), \qquad \text{where } 0 \le \lambda \le 1. \tag{18}$$

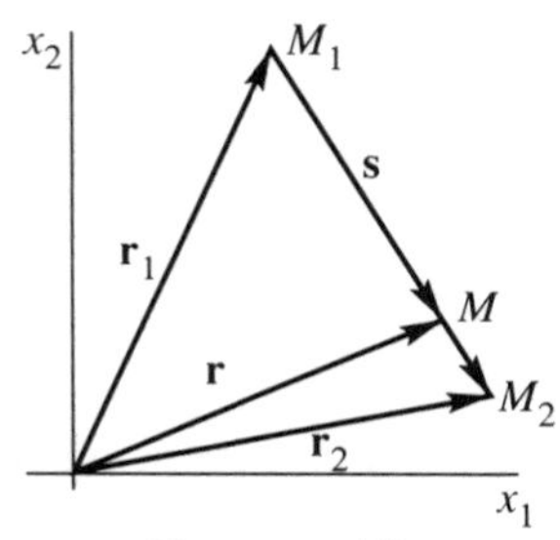

FIGURE 13

Our arguments—and so formula (18)—extend unchanged from plane (2-dimensional case) to space (3-dimensional case). It is natural, therefore, for an n dimensional space with an arbitrary n, to assume that the radius vector $\mathbf{r}$ of an arbitrary point M on the line segment M_1M_2 is found by the same formula (18).

In the plane, each of the points—for example, M_1 and M_2—is defined by two coordinates: $M_1(x_1^{(1)}, x_2^{(2)})$; $M_2(x_1^{(2)}, x_2^{(2)})$. The corresponding vectors have the same coordinates: $\mathbf{r}_1 = \overline{OM}_1 = \{x_1^{(1)}, x_2^{(2)}\}$, and $\mathbf{r}_2 = \overline{OM}_2 = \{x_1^{(2)}, x_2^{(2)}\}$. The coordinates of the vector $\mathbf{r} = \{x_1, x_2\}$ (and of M) are found by the formula (18), which in coordinate form yields two relations:

$$x_1 = (1-\lambda)x_1^{(1)} + \lambda x_1^{(2)}, \qquad x_1 = (1-\lambda)x_2^{(1)} + \lambda x_2^{(2)}.$$

In space $(n = 3)$, yet another relation is added, for the third coordinate:

$$x_3 = (1-\lambda) - x_3^{(1)} + \lambda x_3^{(2)}.$$

In n-dimensional space, points $M_1(x_1^{(1)}, \dots, x_1^{(n)})$ and $M_2(x_1^{(2)}, \dots, x_n^{(2)})$ are defined each by n coordinates, and the coordinates of the point $M(x_1, \dots, x_i, \dots, x_n)$ are determined from n relations

$$x_i = (1-\lambda)x_i^{(1)} + \lambda x_i^{(2)} \qquad (i = 1, \dots, n)$$

or from one vector relation

$$\mathbf{r} = (1-\lambda)\mathbf{r}_1 = \lambda\mathbf{r}_2. \tag{19}$$

We shall call the set of all such points M a *line segment* M_1M_2 in n-dimensional space.

A point set of space T is called a *convex body (figure)*, if, along with any two of its points M_1 and M_2, this set contains all the points of the line segment M_1M_2.

Examples of plane convex bodies are a disc, a sector of a disc, a segment, and any regular polygon. In space, such are a ball and a cone, as well as a pyramid and a prism that have for their bases regular polygons. Figure 14a on the following page shows a convex figure, and 14b shows a figure that is not convex.

By *intersection* of bodies (figures) is meant the set of points belonging to each of the original bodies. The intersection of any number of convex bodies

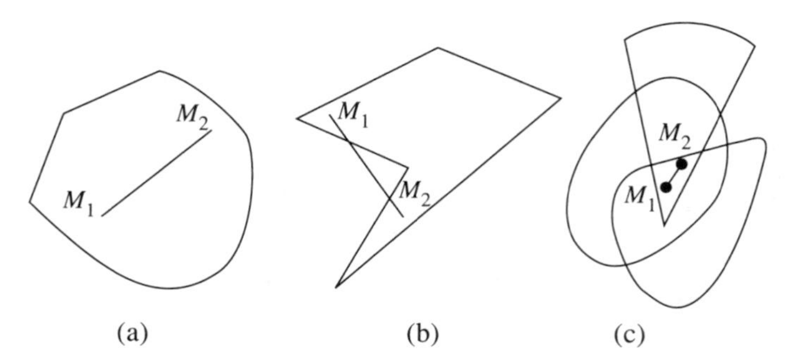

FIGURE 14

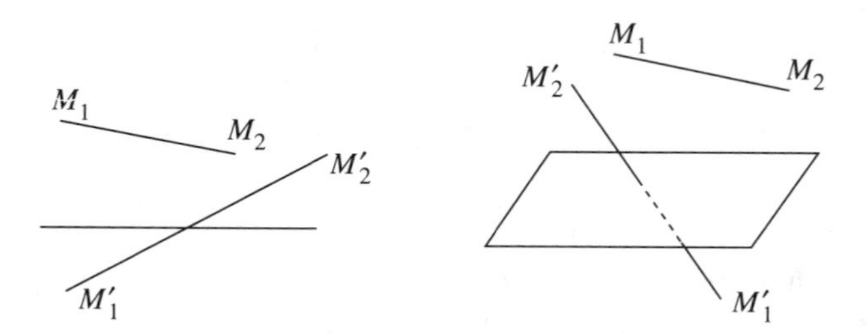

FIGURE 15

is also a convex body. Let us make sure of it (see Figure 14c). Choose any two points M_1 and M_2 in the intersection. These points belong to each of the intersecting sets. But each of them is convex and thus contains the entire line segment M_1M_2. By virtue of this, M_1M_2 wholly belongs to their intersection too. This means that the intersection is a convex body.

Let us set in a plane a straight line by an equation $a_0 + a_1x_1 + a_2x_2 = 0$. It divides the entire plane into two half-planes. Further, let us mark in the plane a pair of points, M_1 and M_2.

It is clear that if the points M_1 and M_2 belong to the one half-plane, then the line segment M_1M_2 joining them does not intersect the dividing line. If the segment $M_1'M_2'$ joining the points M_1' and M_2' intersects this line, then the points M_1' and M_2' lie in different half-planes (see Figure 15).

The same is true in three-dimensional space, making it possible to see whether the points lie on one, or on different sides of a given plane $a_0 + a_1x_1 + a_2x_2 + a_3x_3 = 0$. Further, these conditions, unchanged, are extended to the general case of an n-dimensional space. An n-dimensional plane (called *hyperplane*) is defined by a linear equation with respect to $x_1, \dots, x_n$

$$a_0 + a_1x_1 + a_2x_2 + \cdots + a_nx_n = 0. \tag{20}$$

It turns out that a set of points in n-dimensional space T, not lying in the hyperplane (20), can be broken into two parts, so that the following conditions are satisfied: the line segment joining any pair of points from one part does not intersect the hyperplane (20); the line segment joining any pair of points from different parts intersects the hyperplane (20).

Let us check. To shorten the notation, we shall confine ourselves to the plane case ($n = 2$). For any n, the arguments remain the same. Define a straight line

$$a_0 + a_1x_1 + a_2x_2 = 0. \tag{21}$$

The linear form equal to the left-hand side of this equation is

$$\Phi = a_0 + a_1x_1 + a_2x_2. \tag{22}$$

At the points of the plane not lying on the straight line (21), the form is $\Phi \neq 0$. Let us break the plane into two parts and denote them by T^+ and T^-. The part T^+ consists of all points at which $\Phi > 0$, and T^-, of all points at which $\Phi < 0$. Let us now show that our statement holds for both T^+ and T^-.

Let $M_1(x^{(1)}, x_2^{(1)})$ and $M_2(x_1^{(2)}, x_2^{(2)})$ be two points from one and the same part, for example, T^+. Then

$$\Phi(M_1) = a_0 + a_1x_1^{(1)} + a_2x_2^{(1)} > 0, \tag{23}$$

$$\Phi(M_2) = a_0 + a_1x_1^{(2)} + a_2x_2^{(2)} > 0. \tag{24}$$

Multiply (23) and (24), respectively, by the numbers $1 - \lambda > 0$ and $\lambda > 0$ and add the inequalities obtained. We get

$$\begin{aligned}(1-\lambda)[a_0 + a_1x_1^{(1)} + a_2x_2^{(1)}] + \lambda[a_0 + a_1x_1^{(2)} + a_2x_2^{(2)}]\\ = a_0 + a_1[(1-\lambda)x_1^{(1)} + \lambda x_1^{(2)}] + a_2[(1-\lambda)x_2^{(1)} + \lambda x_2^{(2)}] > 0.\end{aligned} \tag{25}$$

But $x_1 = (1-\lambda)x_1^{(1)} + \lambda x_1^{(2)}$ and $x_2 = (1-\lambda)x_2^{(1)} + \lambda x_2^{(2)}$, given $0 \leq \lambda \leq 1$, are the coordinates of any point M on the line segment M_1M_2. The inequality (25) can be written as $\Phi(M) = (1-\lambda)\Phi(M_1) + \lambda\Phi(M_2)$. The result $\Phi(M) > 0$ means that M—and hence the entire line segment M_1M_2—belong to T^+. Now assume, for example, that $\Phi(M_1) > 0$ and $\Phi(M_2) < 0$, i.e., that M_1 and M_2 belong to different parts. M will be on the straight line (21) if $\Phi(M) = 0$, i.e., at $(1-\lambda)\Phi(M_1) + \lambda\Phi(M_2) = 0$. Hence we find

$$\lambda = \Phi(M_1)/[\Phi(M_1) - \Phi(M_2)]. \tag{26}$$

But $\Phi(M_1) > 0$, $\Phi(M_2) < 0$; therefore, λ, found from (26), satisfies the inequalities $0 \leq \lambda \leq 1$ and thus determines a point M of the line segment M_1M_2, which simultaneously belongs to the straight line (21). This proves that M_1M_2 intersects the straight line (21).

Thus any straight line (21) divides the entire plane T into two parts or half-planes. Each of them is bounded by the straight line (21), which we shall agree to include in each of the half-planes.

If we now turn to the hyperplane (20), then, transferring the arguments to this general case, we can see the following:

The hyperplane (20) divides the space into two parts of half-planes. In each of the half-spaces, the form $\Phi = a_0 + a_1x_1 + \cdots + a_nx_n$ retains its sign. In different half-spaces, Φ has a different sign. Therefore, any hyperplane

$a_0 + a_1x_1 + \cdots + a_nx_n = C$, where $C = \text{const}$, lies entirely either in one or the other half-space, depending on the sign of C.

Let us show that a half-plane (half-space), bounded by any straight line (21) (hyperplane (20)), is a convex body. We have already seen that if the points M_1 and M_2 belong to the same half-plane (half-space), then $\Phi(M_1)$ and $\Phi(M_2)$ have identical signs. $\Phi(M)$ has the same sign for any point M of the segment M_1M_2. This means that the entire segment M_1M_2 belongs to the same half-plane (half-space).

Now let us consider an arbitrary linear inequality with respect to the variables x_1 and x_2. It can always be written as

$$a_0 + a_1x_1 + a_2x_2 \geq 0. \tag{27}$$

We shall call the set of plane points, whose coordinates satisfy this inequality, the set of solutions of this inequality.

To make sure that the set of solutions of a linear inequality is a half-plane, we shall pass from the inequality (27) to the equality $a_0 + a_1x_1 + a_2x_2 = 0$. This equation determines a straight line in the plane, which, as we know, divides the plane into two half-planes. At each point of one of the half-planes, $F = a_0 + a_1x + a_2x_2 \geq 0$. This half-plane is the set of solutions for the inequality (27).

For example, let us consider the inequality

$$2 - x_1 - x_2 \geq 0.$$

The straight line $2 - x_1 - x_2 = 0$ divides the plane into two parts. The set of solutions of the inequality (I) is the half-plane to which belongs the origin of the coordinates. The other half-plane is the set of solutions of the inequality $2 - x_1 - x_2 \leq 0$.

By analogy with the above, the set of solutions of the linear inequality

$$a_0 + a_1x_1 + a_2x_2 + \cdots + a_nx_n \geq 0$$

with respect to n variables $x_1, \ldots, x_n$ is a half-space of n-dimensional space.

It is natural to define as a *set of solutions of a system of linear inequalities* a set of points in space whose coordinates satisfy each of the inequalities of the system.

Since each inequality in a system determines a certain half-space, the set of solutions for the system of inequalities

$$\begin{gathered} a_{10} + a_{11}x_1 + a_{12}x_2 + \cdots + a_{1n}x_n \geq 0, \\ a_{20} + a_{21}x_1 + a_{22}x_2 + \cdots + a_{2n}x_n \geq 0, \\ \cdots\cdots\cdots\cdots\cdots\cdots\cdots\cdots\cdots \\ a_{m0} + a_{m1}x_1 + a_{m2}x_2 + \cdots + a_{mn}x_n \geq 0 \end{gathered}$$

is the intersection of m half-spaces. This intersection is the empty set, if the system of inequalities is inconsistent.

The intersection (if it is not empty) of any number of half-spaces is called a *convex polyhedron* (or a *polyhedral convex set*, if it is not bounded).

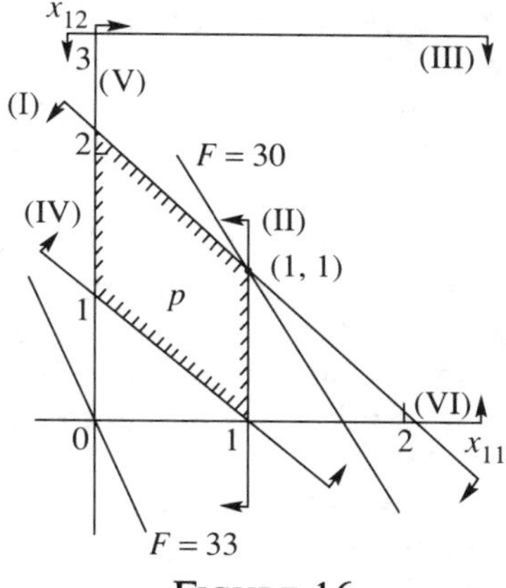

FIGURE 16

Thus, the set of solutions of a consistent system of linear inequalities is a convex polyhedron (or an unbounded polyhedral convex set). The polyhedron is bounded by hyperplanes whose equations are obtained when equality signs are substituted for the inequality signs in the system.

To illustrate, we now return to our example from the previous section; we shall consider again the system of constraints (I)–(VI) in that problem and will solve it geometrically.

This system of inequalities determines a convex polygon P, lying in the first quarter and represented in Figure 16.

Among the points of the polygon P we have to find those at which the form $S(X)$ from (15) takes on the least possible value. To this end, let us consider the level lines of the form $S(X)$. For $S(X) = 33$, the level line has the equation $33 = 33 - 2x_{11} - x_{12}$, i.e., $x_{12} = -2x_{11}$.

This straight line passes through the origin with a slope $k = -2$ and is shown in the figure by a corresponding line. Every other level line $S(X) = C$ is parallel to this line.

Let us find among the level lines of $S(X)$ the one that corresponds to the least of the possible values and simultaneously intersects the polygon P. It is the line (shown in Figure 16) that passes through a vertex of P—the point $(1, 1)$. It corresponds to the value of $S(X) = 33 - 2 - 1 = 30$. Thus, the minimum cost $\min S(X) = 30$ and is determined by the values of the variables

$$x_{11} = 1, \qquad x_{12} = 1, \qquad x_{13} = 0,$$
$$x_{22} = 2, \qquad x_{23} = 1, \qquad x_{21} = 0.$$

This means that the optimal plan is to send from the center A_1 one player each to the clubs B_1 and B_2. From the center A_2, two players are transferred to the club B_2, and one to the club B_3.

Before formulating the common characteristics of solutions to problems like the one above (or the *assignment problem*), let us analyze yet another example which, as you will see, contains a new element, namely, the set of feasible solutions in this example is not a convex polyhedron (or polygon, as in the just solved problem) but an unbounded polyhedral convex set.

6.4. The problem of an athlete's diet. One of the first to be solved by linear programming method was the following problem, consider independently by two American mathematicians—Stigner in 1945 and Koopmans in 1947.

Here it is. To keep in good health and maintain the ability to work, one should of course consume a certain daily amount of nutrients: proteins, fats, carbohydrates, vitamins, and so on. Special requirements (for example, a limited consumption of fats and carbohydrates) are set for those athletes. These requirements are based on the advice of specialists (physicians, coaches). However, it is not athletes alone who need a rational diet. Therefore, the following arguments may well be of general interest.

The amount of nutrients $\beta_1, \beta_2, \dots, \beta_m$ in the different foods $\pi_1, \pi_2, \dots, \pi_n$ varies. Let us denote by a_{ij} the amount (in certain units) of the nutrient β_i in the food π_j. From the variables a_{ij}, we can compose a matrix $A = (a_i)$ of m rows and n columns.

Let us further assume that the cost of a unit of the fold π_j is c_j $(j = 1, \dots, n)$, and the minimal, say daily, allowance of the nutrient β_i is expressed by the number b_i $(i = 1, \dots, m)$.

Let us denote by x_j $(j = 1, \dots, n)$ the amount of the food π_j acquired for the diet (it is implied that the entire food purchased of consumed). In this case, the total amount of the nutrient β_i in all the foodstuffs will be

$$a_{i1}x_1 + \cdots + a_{ij}x_j + \cdots + a_{in}x_n .$$

This amount should not be less than the minimum norm b_i, which leads to m inequalities

$$\sum_{j=1}^{n} a_{ij}x_j \geq b_i \qquad (i = 1, \dots, m). \tag{28}$$

The total cost of the food purchased is

$$F(X) = \sum_{j=1}^{n} c_j x_j . \tag{29}$$

Naturally, none of the values x_j can be negative, i.e., $x_j \geq 0$, since the food π_j is either bought in the amount x_j of dropped from the diet.

Thus, the real-world problem of finding the least expensive but full sufficient diet leads us to the following mathematical problem (model).

We are given a system (28) of m linear inequalities in n unknowns $x_i, \dots, x_n$. We have to find among the nonnegative solutions of the system (28) one that gives the linear form (29) of these variables the minimum value (minimizes the form $F(X)$).

This is a typical linear programming problem, set in the *standard form* (the constraints (28) are inequalities). Let us use the vector notation. Introduce, besides the matrix A, column vectors of independent variables and constant terms

$$A = \begin{pmatrix} a_{11} & \cdots & a_{1n} \\ \cdots & \cdots & \cdots \\ a_{m1} & \cdots & a_{mn} \end{pmatrix}, \qquad X = \begin{pmatrix} x_1 \\ \cdots \\ x_n \end{pmatrix}, \qquad B = \begin{pmatrix} b_1 \\ \cdots \\ b_m \end{pmatrix},$$

as well as a row vector $C = (c_i, \dots, c_n)$ of the coefficients of the form $F(X)$.

Applying the rule of matrix multiplication to the matrix A and the column vector X, as well as to the row vector C and the column B, we can rewrite briefly the system of constraints and the form $F(X)$ as $AX \geq B$, $F(X) = CX$, and the requirement of nonnegativity as $X \geq 0$.

Linear programming offers universal methods for solving the above and mathematically similar problems, including those in which all constraints are exact equalities (as in the soccer club problem or the assignment problem).

The first of such methods—the method of solution factors—was developed by L. V. Kantorovich in 1939 and improved by him and M. K. Gavurin in 1940. In 1949, the first work on the general problems in linear programming was published in the United States. In it, J. Dantzig described the solution of the general linear programming problem by the simplex method which has since gained wide recognition. Setting the simplex method aside for the time being, let us try to understand the mathematical model of the rationing problem and perform the necessary analysis. We shall confine ourselves to the simplest alternative, with just five nutrients ($m = 5$) and two foods ($n = 2$).

Let us give all known variables (a_{ij} and b_i) numerical values. They are shown in Table 8 and are purely illustrative.

TABLE 8

Nutrients	Foods		Ration
	π_1	π_2	
β_1	1	5	10
β_2	3	2	12
β_3	2	4	16
β_4	2	2	10
β_5	1	0	1

Let the (per unit) cost of π_1 and π_2 be 2 and 3 monetary units, respectively. The constraints (28), conditions of nonnegativity of variables, and the form to be minimized will be

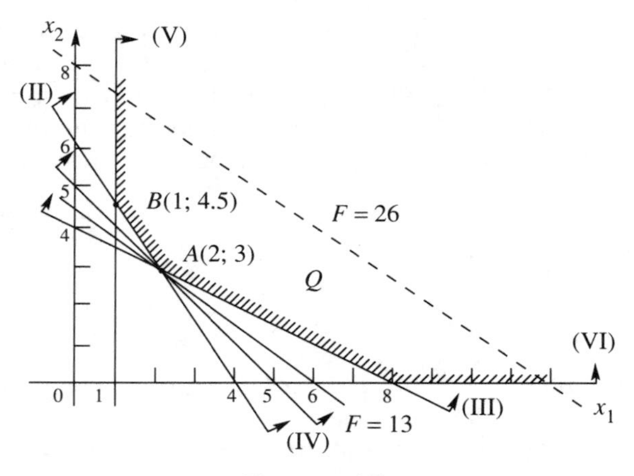

FIGURE 17

$$
\begin{array}{ll}
\text{(I)} & x_1 + 5x_2 \geq 10, \\
\text{(II)} & 3x_1 + 2x_2 \geq 20, \\
\text{(III)} & 2x_1 + 4x_2 \geq 16, \\
\text{(IV)} & 2x_1 + 2x_2 \geq 10, \\
\text{(V)} & x_1 \geq 1, \\
\text{(VI)} & x_2 \geq 0,
\end{array}
$$

$$F(X) = 2x_1 + 3x_2$$

(as it is a corollary of the inequality $x \geq 1$, inequality $x \geq 0$ is not included in this system).

Figure 17 shows the region Q of feasible solutions, defined by the system of linear inequalities (I)–(VI), and the level lines of the form to be minimized, F.

We can see that the least value is attained by the form $F(X)$ at the point $A(2, 3)$, which is one of the vertices of Q. Hence, the least expensive diet costs $F(A) = 13$ monetary units, expended on purchasing the foods π_1 and π_2 in the amounts $x_1 = 2$ and $x_2 = 3$, respectively.

Note that in this example the region Q of feasible solutions extends upward indefinitely. So the form $F(X)$ can reach arbitrarily large values on Q.

And that means that the diet may be as costly as you please.

Let us change the cost of the foodstuffs π_1 and π_2, assuming them to be equal to 3 and 2 monetary units, respectively. In that case, it is necessary to minimize $F(X) = 3x_1 + 2x_2$. The level lines of this form are parallel to the side AB of Q. Hence, the least value of $F(X)$ is attained at every point of AB.

6.5. Linear programming problems. We have examined the geometric method of solving two linear programming problems. One is the soccer club problem; it is in *canonical form*, all constraints (except that of nonnegativity) being equations. The other is the diet problem; it is in *standard form*, all constraints being inequalities.

In both examples we saw that if an optimal solution exists and is single valued, it is achieved at a vertex (corner point) of a polygon (or a polyhedral unbounded set) of feasible solutions. Therefore, once there is an optimal solution, there will always be at least one vertex at which this solution is reached.

This is a general inference that holds for every linear programming problem. It is one of the basic facts of the general theory, and we speak of it later on (see 6.7).

In the above problems, the number of slack variables, by which the rest of the variables are expressed, equals two. That exactly is what made it possible to construct a set of feasible solutions as a polygon (a polygonal set) in the plane. If the number of slack variables is three, one has to pass to three-dimensional space, where the set of feasible solutions is a polyhedron (or an unbounded polyhedral set). In this case too, one can try in principle to solve the problem geometrically. Practically speaking, however, it is a hopeless proposition even in the least complex of cases.

If the number of slack variables is greater than three, the geometric visualization vanishes, and one has to resort to universal methods, namely, successive plan improvement (the simplex method), the dual simplex method (revised estimation method), best-fit method, and so on [**6**].

The simplex method is adapted to solution of linear programming problems in canonical form.

Let us formulate such a problem. We are given a system

$$\begin{aligned} a_{11}x_1 + a_{12}x_2 + \cdots + a_{1n}x_n &= b_1, \\ \cdots\cdots&\cdots\cdots \\ a_{m1}x_1 + a_{m2}x_2 + \cdots + a_{mn}x_n &= b_m \end{aligned} \tag{30}$$

of m linear equations in n unknowns $x_1, \dots, x_n$ and a linear form

$$F(X) = c_0 + c_1x_1 + \cdots + c_nx_n \tag{31}$$

with respect to the same unknowns. We have to find among all nonnegative $x_j \geq 0$ $(j = 1, \dots, n)$ solutions of system (30) those that minimize the form (31).

System (30) is the *system of constraints of the problem*. Any nonnegative solution $X = (x_1, \dots, x_n)$, $x_j \geq 0$ $(j = 1, \dots, n)$ is called a *feasible solution* or *schedule*.

Note that the problem of maximizing the form $F(X)$—i.e., finding among all feasible solutions of system (30) those that give the form (31) its greatest value—reduces to the problem of minimizing the form $\Phi(X) = -F(X)$. Indeed, the minimum value of $\Phi(X)$ equals the maximum value of $F(X)$ taken with the opposite sign.

The reader must have noticed the linear inequalities among the constraints in the foregoing problems. Let us now show that by introducing additional unknowns one can pass from inequality to equality constraints. Indeed, let the inequality

$$\alpha_1 x_1 + \cdots + \alpha_n x_n + \beta \geq 0 \tag{32}$$

be one of the problem's constraints. Let us introduce another unknown (denoted by x_{n+1}) with the help of the equation

$$x_{n+1} = \alpha_1 x_1 + \cdots + \alpha_n x_n + \beta\,. \tag{33}$$

Clearly, the condition $x_{n+1} \geq 0$ of nonnegativity of x_{n+1} is equivalent to satisfying inequality (32). This means that if the set $x_1^0, \ldots, x_n^0, x_{n+1}^0$ of nonnegative values of the variables $x_1, \ldots, x_n, x_{n+1}$ satisfies equation (33), then it satisfies inequality (32) as well.

The converse is also true: if the nonnegative variables $x_1^0, \ldots, x_n^0$ satisfy inequality (32), then $x_{n+1}^0 = \alpha_1 x_1^0 + \cdots + \alpha_n x_n^0 + \beta$, found from equation (33), is also nonnegative.

This proves that inequality (32) and equality (33) are equivalent. In this way, any other inequality constraint can also be replaced by an equivalent equality constraint. As the result (although the number of unknowns will increase), the system of constraints will be in the form of (30).

And now, we present a few general remarks with respect to system (30).

The proof of the consistency criterion for system (30)—the Kronecker-Capelli theorem—is set forth in any course of linear algebra. According to this theorem (see **[6]**), system (30) is consistent if and only if the ranks of the matrix $A = (a_{ij})$ of system (30) and of the argmented matrix (obtained by joining to A the column of the free terms) coincide.

Let $r = n$; then the solution of system (30) is unique and may be found, for example, by Cramer's rule or the Gauss method. If the values x_j^0 $(j = 1, \ldots, n)$ of all the unknowns found are nonnegative, then the solution $X^0 = (x_1^0, \ldots, x_n^0)$ is, indeed, the optimal one (there are no other solutions). If, however, there is among x_j^0 at least one that is negative, then the problem has no solution. Hence, the $r < n$ case is of interest. In this case, as is known from the course of linear algebra, r unknowns (called *basic*) are linearly expressed through the rest $k = n - r$ of unknowns (called *slack*). It is convenient to renumber the unknowns denoting the slack ones by $x_1, x_2, \ldots, x_k$ and the basic ones by $x_{k+1}, x_{k+2}, \ldots, x_{k+r}$ $(k+r = n)$. Thus, from system (30) we pass to an equivalent system

$$\begin{aligned} x_{k+1} &= \beta_1 + \alpha_{11}x_1 + \alpha_{12}x_2 + \cdots + \alpha_{1k}x_k\,,\\ x_{k+2} &= \beta_2 + \alpha_{21}x_1 + \alpha_{22}x_2 + \cdots + \alpha_{2k}x_k\,,\\ &\cdots\cdots\cdots\cdots\cdots\cdots\cdots\cdots\cdots\\ x_{x+r} &= \beta_r + \alpha_{r1}x_1 + \alpha_{r2}x_2 + \cdots + \alpha_{rk}x_k\,. \end{aligned} \tag{34}$$

Now we can also express the form $F(X)$ only through slack variables, replacing the basic variables in (32) according to formulas (34). The result will be

$$F(X) = \gamma_0 + \gamma_1 x_1 + \cdots + \gamma_k x_k\,, \tag{35}$$

where $\gamma_0, \ldots, \gamma_k$ are coefficients.

Our linear programming problem requires that only feasible (nonnegative) values of the variables be taken into account. Therefore, the inequalities $x_j \geq 0$ $(j = 1, \ldots, n)$ must be satisfied, or, in a more detailed notation,

$$\begin{aligned} &x_1 \geq 0, \ x_2 \geq 0, \ldots, x_k \geq 0, \\ &\beta_1 + \alpha_{11}x_1 + \cdots + \alpha_{1k}x_k \geq 0, \\ &\cdots\cdots\cdots\cdots\cdots\cdots\cdots\cdots \\ &\beta_r + \alpha_{r1}x_1 + \cdots + \alpha_{rk}x_k \geq 0. \end{aligned} \tag{36}$$

Thus, starting from system (30) of linear equalities with respect to n unknowns, we arrive at a system of n linear inequalities with respect to k unknowns. This gives rise to the following mathematical problem.

Given a system (36) of n linear inequalities with respect to k unknowns $x_1, \ldots, x_k$ and a linear form (35) with respect to the same unknowns, find among all the solutions of (36) those that give the form (35) the least value.

It remains to recall that this is also a linear programming problem, but in standard form (all constraints are inequalities). Thus, a canonical linear programming problem can be restated in standard form by expressing the basic variables (including the form to be minimized) through slack variables. One can pass from standard to canonical representation by introducing additional variables in the way indicated above (see p. 88).

So two ways of giving a geometric interpretation to a linear programming problem present themselves. It is possible to remain in the n-dimensional space $R^{(n)}$ of variables $x_1, x_2, \ldots, x_n$ or to pass to the k-dimensional space $R^{(k)}$ $(k < n)$ of slack variables $x_1, x_2, \ldots, x_k$.

All is ready for the latter interpretation. Indeed, we already know that each of the inequalities (36) defines in $R^{(k)}$ a certain half space, and the set of all inequalities defines a convex polyhedron (or a convex polyhedral set) Q (see 6.6). The first k inequalities of system (36) indicate that the polyhedron lies in the positive orthant.

Further, all possible hyperplanes $\gamma_0 + \gamma_1 x_1 + \gamma_2 x_2 + \cdots + \gamma_n x_n = C$ of equal values of the form $F(X)$, intersecting the polyhedron Q of feasible solutions are examined, and the points corresponding to the least value of C are singled out. It is proved further on (6.7) that among these points there is always at least one *vertex* (see below for an exact definition) of the polyhedron Q.

We turn now to the geometric interpretation of a linear programming problem in the space $R^{(n)}$.

Let, for example, $X^{(1)} = (x_1^{(1)}, \ldots, x_n^{(1)})$ and $X^{(2)} = (x_1^{(2)}, \ldots, x_n^{(2)})$ be any two feasible solutions of constraint system (30). It means that, substituting these solutions into system (30), two identities will occur. Let us write them down in matrix form:

$$AX^{(1)} = B, \qquad AX^{(2)} = B. \tag{37}$$

Let us put together a linear combination

$$X = (1 - \lambda)X^{(1)} + \lambda X^{(2)} \tag{38}$$

of these solutions with coefficients $\alpha_1 = 1 - \lambda$, $\alpha_2 = \lambda$, given an arbitrary value of λ between zero and one $(0 \le \lambda \le 1)$. Such a linear combination is called *convex*.

If we compare (38) and formula (19) for the radius vector $\mathbf{r}$ of an arbitrary point on the line segment whose ends are determined by the vectors $\mathbf{r}_1$ and $\mathbf{r}_2$, we shall see that they are identical. Nor is it fortuitous: each of the solutions $X^{(1)}$ and $X^{(2)}$ are n-dimensional column vectors, and the end of the vector X passes (for $0 \le \lambda \le 1$) through the entire line segment connecting $X^{(1)}$ and $X^{(2)}$.

By substituting it into (30), we shall see that X also is a solution to this system:

$$AX = A(\lambda X^{(2)} + (1-\lambda)X^{(1)}) = \lambda AX^{(2)} + (1-\lambda)AX^{(1)}.$$

But from (37),

$$AX = \lambda B + (1-\lambda)B = B.$$

Furthermore, X is a feasible solution $(X \ge 0)$, since for each of the coordinates it is true that $x_i = \lambda x_i^{(2)} + (1-\lambda)x_i^{(1)} \ge 0$ $(i = 1, \dots, n)$ owing to the fact that $x_i^{(1)} \ge 0$, $x_i^{(2)} \ge 0$, $\lambda > 0$, $1 - \lambda \ge 0$. Thus, it is proved that the set V of all feasible solutions of a linear programming problem in the space $R^{(n)}$ is convex.

6.6. Corner points and convex combinations. A vector X, which belongs to the convex set V, is called an *extreme* or *corner point* or *vertex* of V if for any vectors $X^{(1)}$ and $X^{(2)}$ from V, given $0 \le \lambda \le 1$, it follows from $X = (1-\lambda)X^{(2)} + \lambda X^{(1)}$ that $X = X^{(1)} = X^{(2)}$. In other words, X is the extreme (corner) point of the set V, if X cannot be represented as a convex combination of any two other points out of V, different from X itself.

Geometrically this means that an extreme point cannot lie inside any line segment belonging to V.

The extreme points of a line segment are its ends; those of a triangle are its vertices; those of a pyramid are the points of intersection of its edges. Each boundary point of a circular region is an extreme point. Note that a nonclosed region (for example, the interior of a circle) may have no extreme points.

Let us apply the above definition of a convex combination of two vectors (points) to the general case. A vector X from $R^{(n)}$ is called a *convex combination* of vectors $X^{(1)}, \dots, X^{(l)}$, if it can be represented as a linear combination

$$X = \lambda_1 X^{(1)} + \lambda_2 X^{(2)} + \cdots + \lambda_l X^{(l)} \tag{39}$$

with nonnegative coefficients $\lambda_1, \lambda_2, \dots, \lambda_l$ adding up to one:

$$\lambda_1 + \lambda_2 + \cdots + \lambda_n = 1.$$

Thus, for example, the linear combination

$$\tfrac{1}{4}X^{(1)} + \tfrac{2}{3}X^{(2)} + \tfrac{1}{12}X^{(3)} \qquad (\tfrac{1}{4} + \tfrac{2}{3} + \tfrac{1}{12} = 1)$$

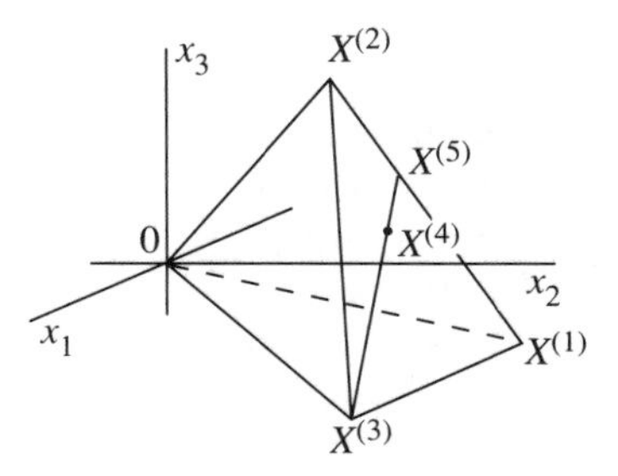

FIGURE 18

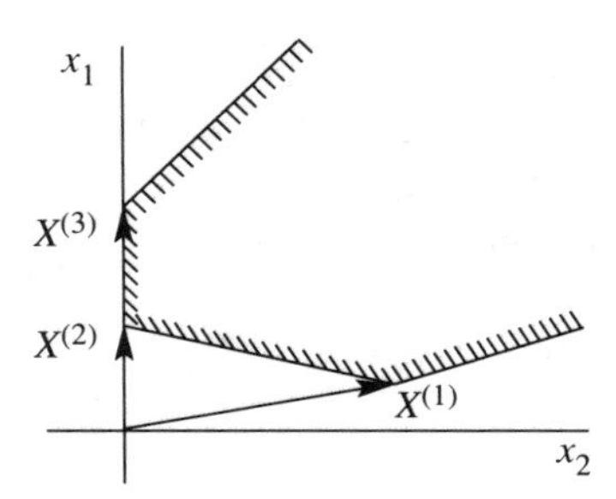

FIGURE 19

is convex. However, the following combination is not convex

$$\tfrac{1}{4}X^{(1)} + \tfrac{2}{3}X^{(2)} + \tfrac{1}{6}X^{(3)} \qquad (\tfrac{1}{4} + \tfrac{2}{3} + \tfrac{1}{6} \neq 1),$$

nor is

$$\tfrac{1}{3}X^{(1)} + X^{(2)} - \tfrac{1}{3}X^{(3)} \qquad (\lambda_3 = -\tfrac{1}{3} < 0).$$

Let us plot in $R^{(n)}$ vectors $X^{(1)}, \dots, X^{(l)}$ and look at their convex combinations. The set of all such vectors (39) is called the *convex hull* of $X^{(1)}, \dots, X^{(l)}$.

For example, the convex hull of two vectors is the line segment $X = (1-\lambda)X^{(1)} + \lambda X^{(2)}$; that of three vectors (not belonging to the same line segment) is a triangle; that of four noncoplanar vectors is a tetrahedron (see Figure 18).

In the general case, the convex hull of a finite number of vectors $X^{(1)}, \dots, X^{(m)}$ is called a *convex polyhedron* (if the convex hull is bounded) or a *convex polyhedral set* (if it is not bounded) (Figure 19). For example, the convex hull of the vectors $X^{(1)} = (1, 0, 0)$, $X^{(2)} = (0, 1, 0)$, $X^{(3)} = (0, 0, 1)$, $X^0 = (0, 0, 0)$ is a tetrahedron with vertices at these points. It can also be defined by the linear inequalities $x_1 \geq 0$, $x_2 \geq 0$, $x_3 \geq 0$, $x_1 + x_2 + x_3 \leq 1$.

This is not accidental. The German mathematician Hermann Weil proved that both definitions of the convex polyhedron (as a nonempty intersection of a finite number of half spaces or as the convex hull of a finite number of vectors) are equivalent.

It can be verified that a convex hull itself is a convex set and that every convex set coincides with its convex hull. With reference to a convex polyhedron this means that each of its points X is a convex combination of the corner points $X^{(1)}, \dots, X^{(m)}$, and any convex combination of corner points is a point of the polyhedron.

Briefly, a convex polyhedron coincides with the convex hull of all of its extreme points:

$$X = \sum_{i=1}^{m} \lambda_1 X^{(i)}, \qquad \lambda_i \geq 0\ (i = 1, \dots, m),\ \sum_{i=1}^{m} \lambda_i = 1.$$

For example, the point $X^{(4)}$ (Figure 18) belongs to the line segment defined by the vectors $X^{(3)}$ and $X^{(5)}$, and therefore $X^{(4)} = \lambda X^{(3)} + (1-\lambda)X^{(5)}$ $(0 \leq \lambda \leq 1)$. For a similar reason, $X^{(5)} = \mu X^{(1)} + (1-\mu)X^{(2)}$ $(0 \leq \mu \leq 1)$. Hence,

$$\begin{aligned} X^{(4)} &= \lambda X^{(3)} + (1-\lambda)[\mu X^{(1)} + (1-\mu)X^{(2)}] \\ &= \lambda_1 X^{(1)} + \lambda_2 X^{(2)} + \lambda_3 X^{(3)}, \end{aligned}$$

where $\lambda_1 = \lambda$, $\lambda_2 = (1-\lambda)\mu$, $\lambda_3 = (1-\lambda)(1-\mu)$ are nonnegative and

$$\lambda_1 + \lambda_2 + \lambda_3 = \lambda + (1-\lambda)\mu + (1-\lambda)(1-\mu) = 1.$$

The latter means that $X^{(4)}$ (an arbitrary point of the triangular face) is a convex combination of the corner points $X^{(1)}$, $X^{(2)}$, $X^{(3)}$.

Let us examine in $R^{(n)}$ $n+1$ points $X^{(j)} = (a_{1j}, a_{2j}, \dots, a_{nj})$ $(j = 1, \dots, n+1)$. These points are said to be in *general position* if the determinant

$$\begin{bmatrix} 1 & 1 & \cdots & 1 \\ a_{11} & a_{12} & \cdots & a_{1,n+1} \\ a_{21} & a_{22} & \cdots & a_{2,n+1} \\ \cdots & \cdots & \cdots & \cdots \\ a_{n1} & a_{n2} & \cdots & a_{n,n+1} \end{bmatrix}$$

is not equal to zero. The convex hull of $n+1$ points in general position in n-dimensional space is called an *n-dimensional simplex.* Thus, a zero-dimensional simplex is a point, a one-dimensional simplex is a line segment, a two-dimensional simplex is a triangle, and a three-dimensional simplex is a tetrahedron.

In a sense, the n-dimensional simplex is the simplest polyhedron in $R^{(n)}$. A convex polyhedron V (although, generally speaking, it is not a simplex) is a set of points of $R^{(n)}$ which are feasible solutions of a linear programming problem. For these reasons, one of the principal methods of searching for the optimal value of a linear form in a set V has been named the *simplex method* (see 6.8).

6.7. Corner points and feasible solutions. Then, why should the corner points of a convex polyhedron enjoy such close attention? The answer lies in the following highly important fact (which we promised to prove).

Any linear form $F(X) = \gamma_0 + \gamma_1 x_1 + \cdots + \gamma_n x_n$, defined on a convex polyhedron Q, attains its minimum (maximum) value at one of the polyhedron's corner points. If $F(X)$ takes on the minimum (maximum) value at more than one corner point, it takes on the same value at each point that is their convex combination.

As agreed, the values of the form $F(X)$ are examined at the points of the polyhedron Q. Let us denote by $X^{(1)}, \ldots, X^{(n)}$ its corner points (a convex polyhedron has a finite number of them by definition). Assume that $F(X)$ reaches its minimum value at a point X^0 of Q; in other words, $F(X^0) = \min F(X) = F^*$. Then, for any other point X from Q, the inequality $F(X) \geq F(X^0)$ is true. If X^0 is one of the corner points, then the first part of the assertion is proved. Suppose that X^0 is not a corner point. But then X^0 is a convex combination of corner points:

$$X^0 = \lambda_1 X^{(1)} + \cdots + \lambda_m X^{(m)}$$
$$\lambda_k \geq 0 \ (k = 1, \ldots, m), \ \lambda_1 + \cdots + \lambda_m = 1.$$

Because of the linearity of the form $F(X)$, its value at X^0

$$\begin{aligned} F(X^0) &= F(\lambda_1 X^{(1)} + \cdots + \lambda_m X^{(m)}) = F(\lambda_1 X^{(1)}) + \cdots + F(\lambda_m X^{(m)}) \\ &= \lambda_1 F(X^{(1)}) + \cdots + \lambda_m F(X^{(m)}) \end{aligned} \tag{40}$$

is a convex combination of values at corner points.

Let us choose the least of the values $F(X^{(1)}), \ldots, F(X^{(m)})$; let it be $F(X^{(k)})$. Then $F(X^{(i)}) \geq F(X^{(k)})$ $(i = 1, \ldots, m)$, and from (40) ensures the inequality

$$F(X^0) \geq F(X^{(k)})(\lambda_1 + \cdots + \lambda_m) = F(X^{(k)}).$$

However, according to our assumption, $F(X^0) \leq F(X)$ for any point X in Q and of course $F(X^0) \leq F(X^{(k)})$. It remains to assume the equality $F(X^0) = F(X)$, i.e., that the minimum value is achieved at least at the corner point $X^{(k)}$. Thus the first part of the assertion is proved.

Let us now prove the second part. Let $F(X)$ achieve the minimum value F^* at each of the corner points $X^{(1)}, \ldots, X^{(l)}$. Consider their arbitrary convex combination.

$$X = \lambda_1 X^{(1)} + \cdots + \lambda_l X^{(l)};$$
$$\lambda_i \geq 0 \ (i = 1, \ldots, l), \ \lambda_1 + \cdots + \lambda_l = 1.$$

Then the value of $F(X)$ at X also is minimal:

$$F(X) = \lambda_1 F(X^{(1)}) + \cdots + \lambda_l F(X^{(l)}) = \lambda_1 F^* + \cdots + \lambda_l F^* = F^*.$$

The assertion is completely proved.

We arrive at the conclusion that the optimal values of the form $F(X)$ on the polyhedron Q of feasible solutions of the constraint system (30) of a linear programming problem should be sought among the values at the corner points of Q.

But how does one find these corner points? Can they be detected, by an examination of system (30)?

Yes, they can. Corner points may be detected when basic feasible solutions to system (30) are found. A more exact answer is given by the following statement.

Each basic feasible solution of the constraint system (30) of a linear programming problem corresponds to a unique corner point (for example, the solution $x_1 = \cdots = x_k = 0$, $x_{k+1} = \beta_1, \ldots, x_{k+r} = \beta_r$ of equations (34) corresponds to the unique corner point $X_* = (0, \ldots, 0, \beta_1, \ldots, \beta_r)$ of the polyhedron Q). Conversely, each corner point X_* of Q determines a certain (generally speaking, not unique) basic feasible solution of system (30).

This statement (for proof see **[17]**) makes it possible to find an upper bound of the possible number of corner points of the polyhedron Q: there are no more corner points than there are basic feasible solutions of system (30).

We shall assume that the rank r of system (30) equals the number m of equations. If not, it is possible to retain in the system only r linearly independent equations and discard the rest. It was noted earlier (see p. 98) that for linear programming problems the $r < n$ case is of interest. It is proved in linear algebra that it is possible to solve system (30) with respect to (basic) unknowns $x_{k+1}, \ldots, x_{k+r}$ and express them through the slack unknowns $x_1, \ldots, x_k$ if and only if in the matrix

$$A = \begin{pmatrix} a_{11} & \cdots & a_{1k} & a_{1,k+1} & \cdots & a_{1,k+r} \\ a_{21} & \cdots & a_{2k} & a_{2,k+1} & \cdots & a_{2,k+r} \\ \cdots & \cdots & \cdots & \cdots & \cdots & \cdots \\ a_{r1} & \cdots & a_{rk} & a_{r,k+1} & \cdots & a_{r,k+r} \end{pmatrix}$$

the minor, formed by the last r columns (coefficients of the basic variables), is not zero.

Similarly, it is possible to express any other r of the basic unknowns $x_{\sigma_1}, \ldots, x_{\sigma_r}$ through the rest of the unknowns—the slack variables $x_{\sigma_{r+1}}, \ldots, x_{\sigma_{r+k}}$ only if the minor of the matrix A, composed of the coefficient columns, corresponding to these basic variables, is not zero:

$$\begin{bmatrix} a_{1\sigma_1} & a_{1\sigma_2} & \cdots & a_{1\sigma_r} \\ a_{2\sigma_1} & a_{2\sigma_2} & \cdots & a_{2\sigma_r} \\ \cdots & \cdots & \cdots & \cdots \\ a_{r\sigma_1} & a_{r\sigma_2} & \cdots & a_{r\sigma_r} \end{bmatrix} \neq 0.$$

It is easy to compute that there are as many minors of order r in the matrix A as there can be combinations of n columns taken r at a time, namely:

$$C_r^n = \frac{n(n-1)(n-2)\cdots(n-r+1)}{1 \cdot 2 \cdot 3 \cdots n}.$$

Hence, there are no more than C_r^n different basic solutions—after all, not all minors of order r of the matrix A are different from zero. Besides, not every basic solution is feasible. Therefore, there are no more than C_r^n corner points. If $r = 5$, $n = 10$, then $C_6^{10} = 756$. One can get an idea of the approximate number of corner points from the following table:

$r = 1$	n	corner points
5	10	$3 \cdot 10^2$
10	20	$2 \cdot 10^5$
20	40	10^{11}

Obviously, the number of corner points is large even for relatively small m and n. In practical problems m and n may run into hundreds. So, even though it is theoretically possible to find by exhaustion the corner points that give the form $F(X)$ is optimal value (after all, there is a finite number of them), practically it is unfeasible.

Fortunately, there is a clear-cut procedure for going through the corner points. Using this procedure, one passes from one corner point to the next, at which the value of the form to be minimized decreases. Thus the search is completed quickly enough (on the average, in m steps, m being the number of linearly independent constraints). This procedure is known as the *simplex method*. Here is a brief description of this method.

6.8. The simplex method. Examine a specific linear programming problem given in canonical form: Minimize the linear form

$$F(X) = 3 - x_4 + x_5, \tag{41}$$

subject to the constraints

$$\begin{aligned} x_2 + 2x_4 + 3x_5 - 7 &= 0, \\ x_3 - x_4 - 3x_5 - 2 &= 0, \\ x_1 + x_4 - x_5 - 2 &= 0, \end{aligned} \tag{42}$$

$$x_i \geq 0 \quad (i = 1, \dots, 5). \tag{43}$$

The consistency test (the Kronecker-Capelli theorem) for system (42) is satisfied. Its rank is three, the number of basic variables is also three, the number of slack variables is two. Hence, this problem can be solved geometrically, considering a plane defined by two slack variables, for example, x_4 and x_5. However, we shall reject the geometric approach (hopefully the reader will not) and solve the problem by other means. To this end, we shall express in (42) the basic variables x_1, x_2, and x_3 through the slack variables x_4 and x_5:

$$\begin{aligned} x_1 &= 2 - x_4 - x_5, \\ x_2 &= 7 - 2x_4 - 3x_5, \\ x_3 &= 2 + x_4 + 3x_5. \end{aligned} \tag{44}$$

The form $F(X)$, as is seen from (41), is already expressed through slack variables.

By dint of constraints (43), the least feasible values of the slack variables are those equal to zero: $x_4 = 0$, $x_5 = 0$; then $x_1 = 2$, $x_2 = 7$, $x_3 = 2$. Hence, we obtain a basic feasible solution of system (44): $X_0 = (2, 7, 2, 0, 0)$. From (41) we have the corresponding value of the form $F(X_0) = 3$.

The choice of zero values of the slack variables is, generally speaking, absolutely unjustified. Let us see if we can make $F(X)$ decrease by increasing x_4 and x_5. From (41) we can see that, since the unknown x_5 is contained in it with a positive coefficient, its increase will merely cause an increase in the form. Simultaneously, x_4 is contained in (41) with a negative coefficient, whereby its increase is accompanied by a decrease of $F(X)$. However, there may be an obstacle to an unlimited growth of x_4, since the values of the basic variables x_1, x_2, x_3 change in the process. They may become negative, i.e., unfeasible solutions will occur. Indeed, it follows from the first equation in (44) that the variable x_1 will become negative at x_4 greater than two (we still keep x_5 equal to zero). Simultaneously, from the second and third equations (44) we obtain that as x_4 changes from zero to two, x_2, though diminishing, remains positive and x_3 increases. Hence, it is more appropriate to give x_4 the value $x_4 = 2$ (instead of $x_4 = 0$). Then $x_1 = 0$, $x_2 = 3$, $x_3 = 4$, $x_4 = 2$, and $x_5 = 0$. Hence, we have a new basic feasible solution $X_1 = (0, 3, 4, 2, 0)$, in which x_1 and x_5 play the part of slack variables. We compute that $F(X_1) = 1$. This, expectedly, is less than $F(X_0) = 3$.

But rather than stop here, we shall compute the expressions of the new basic variables x_2, x_3, and x_4 and the form $F(X)$ through a new set x_1, x_5 of slack variables. To this end, we shall express x_4 from the first equation (44) through x_1 and x_5; we shall obtain $x_4 = 2 - x_1 - x_5$. Substituting this expression into the other two equations (44), we find

$$\begin{aligned} x_2 &= 3 + 2x_1 - x_5, \\ x_3 &= 4 - x_1 + 2x_5, \\ x_4 &= 2 - x_1 - x_5, \end{aligned} \tag{45}$$

$$F(X) = 1 + x_1 + 2x_5. \tag{46}$$

Let us now apply similar arguments to form (46). It is obvious that any increase in the slack variables x_1, x_5 over zero will merely cause the form (46) to grow. Therefore, the solution X_1 is optimal and gives the least value to the form $F(X_1) = 1 = \min F(X)$.

Let us consider another problem and find its solution by the simplex method.

6.9. Indoors or out? The following problem is cited purely for illustration. However, it provides a pattern on which mathematical models can be constructed to supply recommendations in situations less far-fetched than the one we are going to consider.

We shall talk of how to divide the time between training sessions held outdoors or in the gym to ensure the greatest efficiency. Let us suppose that we know the following (in certain units):

(1) the average effectiveness (in different sports) of one hour of training outdoors and in the gym;

(2) the load experience by an athlete per hour of training of a specific kind;

(3) the maximum permissible loads in each kind of training.

The data are presented in Table 9.

TABLE 9

The Kind of Practice	Place		Maximum Permissible Load
	Outdoors	Gym	
Load per Training Hour			
Running	2	3	19
Jumping	2	1	13
Weight Lifting	0	3	15
Skiing	3	0	18
Average Effectiveness of One Hour of Training			
	7	5	

Let us denote by x_1 and x_2 the yet-unknown duration (in hours) of training sessions outdoors and in the gym. Their total effectiveness will amount to

$$\Phi(X) = 7x_1 + 5x_2 . \tag{47}$$

The problem is to maximize $\Phi(X)$ subject to the constraints

$$\begin{gathered} 2x_1 + 3x_2 \leq 19, \quad 2x_1 + x_2 \leq 13, \quad 3x_2 \leq 15, \quad 3x_1 \leq 18, \\ x_1 \geq 0, \quad x_2 \geq 0. \end{gathered} \tag{48}$$

These constraints emerge quite naturally. For example, the first of them means that an athlete's total load during an x_1-hour run outdoors and an x_2-hour run in the gym should not exceed the permissible limit. The rest of the constraints (48) are of a similar kind.

We have arrived, again, at a linear programming problem. Unlike the preceding problem, however, it is in standard, rather than canonical, form. But the method of reasoning described earlier is suited to a linear programming problem in canonical form. Let us bear this in mind and pass to the required form by introducing additional unknowns

$$\begin{aligned} x_3 &= 19 - 2x_1 - 3x_2, \qquad & x_4 &= 13 - 2x_1 - x_2, \\ x_5 &= 15 - 3x_2, & x_6 &= 18 - 3x_1 . \end{aligned} \tag{49}$$

As required by the method, from maximization of $\Phi(X)$ we pass to minimization of

$$\Phi_1(X) = -\Phi(X) = -7x_1 - 5x_2 . \tag{50}$$

Now we have a linear programming problem in canonical form, which is to find among the nonnegative solutions $x_j \geq 0$, $j = 1, \dots, 6$, of system (49) those that give the least value of the linear function $\Phi_1(X)$.

We shall take as slack variables x_1 and x_2 since all the other variables and $\Phi_1(X)$ are already expressed through them. We obtain the initial basic feasible solution X_0:

$$x_1 = 0, \quad x_2 = 0, \quad x_3 = 19, \quad x_4 = 13, \quad x_5 = 15, \quad x_6 = 18, \tag{15}$$

for which $\Phi_1(X_0) = 0$.

Expression (50) contains both slack variables with negative coefficients. Therefore, if either of them increases, $\Phi_1(X)$ will decrease. Let us, for example, start to increase x_2 (while preserving the zero value of x_1). In increasing x_2, we must watch lest the values of the basic variables become negative. At $x_2 = 5$ the basic unknown x_5 turns into zero, while the remaining basic variables continue to be positive. Any further growth of x_2 is impossible.

Let us now choose another pair of slack variables, x_1 and x_5. Let us express through them x_2, x_3, x_4, x_6, and $\Phi_1(X)$. We obtain:

$$\begin{aligned} x_2 &= 5 - \tfrac{1}{3}x_5, & x_3 &= 4 - 2x_1 + x_5, \\ x_4 &= 8 - 2x_1 + \tfrac{1}{3}x_5, & x_6 &= 18 - 3x_1, \end{aligned} \tag{52}$$

$$\Phi_1(X) = -25 - 7x_1 + \tfrac{5}{3}x_2. \tag{53}$$

A feasible basic solution X_1, corresponding to such a choice of slack variables, is

$$x_1 = 0, \quad x_2 = 5, \quad x_3 = 4, \quad x_4 = 8, \quad x_5 = 0, \quad x_6 = 18.$$

Given this solution, the value of the form has decreased, becoming equal to $\Phi_1(X_1) = -25$. This value can be decreased again by increasing x_1 which is contained in (53) with a negative coefficient. So too is x_1 in the expressions of the basic unknowns x_3, x_4, x_5 from (52). Therefore, x_1 can be increased only until one of these unknowns turns into zero for the first time. This will occur for x_3 when $x_1 = 2$ (x_4 and x_5 will still be positive). Let us now take x_3 and x_5 as slack variables and express the rest of the unknowns through them. We shall find that

$$\begin{aligned} x_1 &= 2 - \tfrac{1}{2}x_3 + \tfrac{1}{2}x_5, & x_2 &= 5 - \tfrac{1}{3}x_5, \\ x_4 &= 4 - x_3 - \tfrac{2}{3}x_5, & x_6 &= 12 + \tfrac{3}{2}x_3 - \tfrac{2}{3}x_5, \end{aligned}$$

$$\Phi_1(X) = -39 + \tfrac{7}{2}x_3 - \tfrac{11}{6}x_5.$$

Let us write out the corresponding basic feasible solution X_2:

$$x_1 = 2, \quad x_2 = 5, \quad x_3 = 0, \quad x_4 = 4, \quad x_5 = 0, \quad x_6 = 4;$$

the value of the form has decreased again and become equal to $\Phi_1(X_2) = -39$.

Next we again carry out similar reasoning. We increase x_5 to a value $x_5 = 6$ (at which $x_4 = 0$), take as a new set of slack variables x_3 and x_4, and express through them the rest of the unknowns:

$$x_1 = 5 + \tfrac{1}{4}x_3 - \tfrac{3}{4}x_4, \qquad x_2 = 3 - \tfrac{1}{2}x_3 + \tfrac{1}{2}x_4,$$
$$x_5 = 6 + \tfrac{3}{2}x_3 + \tfrac{3}{2}x_4, \qquad x_6 = 3 - \tfrac{3}{4}x_3 + \tfrac{9}{4}x_4,$$
$$\Phi_1(X) = -50 + \tfrac{3}{4}x_3 + \tfrac{11}{4}x_4.$$

Since the slack variables are contained in the last expression for $\Phi_1(X)$ with positive coefficients, their increase will merely cause the form to grow. Therefore, basic feasible solution X_3

$$x_1 = 5, \quad x_2 = 3, \quad x_3 = 0, \quad x_4 = 0, \quad x_5 = 6, \quad x_6 = 3$$

already is optimal. It gives the form the minimum value $\min\Phi_1 = \Phi_1(X_3) = -50$.

Now we have to return to the initial problem. Its statement mentioned only two variables, x_1 and x_2 (the time of training outdoors and in the gym). Their optimal values, $x_1 = 5$, $x_2 = 3$, are contained in the solution X_3. Furthermore, the maximum value of training efficiency $\max\Phi(X) = 50$.

6.10. Some general inferences. In each of the preceding problems the search for an optimal solution started with some initial basic feasible solution X_0 of a problem's constraint system. Next we passed to another basic feasible solution X_1 which gave the minimized form a smaller value than did the solution X_0. For X_1 we passed to the solution X_2, and so on, until the desired result was obtained. This procedure is the gist of the simplex method.

Recall that each feasible basic solution determines a corner point in the convex polyhedron of a problem's solutions. Therefore, it can be said that starting from some initial corner point we proceed by the simplex method to another corner point, with a smaller value of the form $F(X)$ being minimized, and so on. We already know (see 6.7) that if an optimal value of $F(X)$ on the polyhedron of solutions does exist, it must be attained at least at one of its corner points (i.e., at a certain basic feasible solution).

The simplex procedure can be formalized, and all calculations are carried out with the aid of special tables (*simplex tableaus*), which are easy to work with. Each tableau contains the values of the coefficients with which basic variables and the form are expressed through slack variables (i.e., a certain basic feasible solution). Thus a tableau represents a shorter notation of the constraint system and minimized form of the problem. Following a set procedure (as in the previous examples), from a solution written in a tableau we pass to a tableau with another solution, and so on. The reader will find a description of this procedure practically in every book on linear programming (e.g., [**6**, **1**]), so we shall not dwell on it here.

It should be added that modern computer software offers a special program for solving linear programming problems. So one does not even need to know how to do it "by hand".

The reader must have noted that in the linear programming problems just discussed, all basic variables in each of the basic feasible solutions have values other than zero. This feature is known as nondegeneracy, and a linear programming problem is called *nondegenerate* if in each of its feasible basic solutions ALL basic variables are strictly greater than zero.

Courses in linear programming offer the proof of the following *simplex theorem*. Assume that:

(1) a linear programming problem is nondegenerate;

(2) it has at least one basic feasible solution;

(3) the form to be minimized is bounded below on a set of feasible solutions. (This means that any value of the form on a set of feasible solutions is not less than a fixed number.)

Then, there is at least one *optimal* basic solution. This solution can be reached by the simplex method, starting from any initial basic feasible solution.

Note that if the assumption of nondegeneracy is not satisfied, a situation may occur, in which no optimal solution is reached since there are no obstacles to passing from one basic feasible solution to another. This is an extremely rare phenomenon, known as *cycling*. In geometric terms, cycling means that the solution "gets stuck" at some vertex of the polyhedron Q of solutions, to which correspond several different basic feasible solutions. When this happens, one moves from one such solution to another without shifting to another vertex of Q and without any change in the value of $F(X)$.

To start working by the simplex method, one has to have, as we saw, an initial basic feasible solution X_0. In simple cases X_0 can be found by inspecting (as we have done) the problem's constraint system. However, given a great number of equations and unknowns, a different way of determining X_0 is needed. This, too, can be done by the simplex method. Moreover, this method also makes it possible to establish whether the system of constraints in the region of nonnegative values of the unknowns is consistent, i.e., whether the polyhedron Q of feasible solutions is not empty.

Here is the essence of this procedure. Let us rewrite the constraint system (30) of a linear programming problem as

$$b_i - \sum_{j=1}^{n} a_{ij}x_j = 0 \qquad (i = 1, \dots, m). \tag{54}$$

Without any loss of generality, we can assume that all $b_i \geq 0$ (if $b_i < 0$, then we multiply the ith equation by -1). Introduce (according to the number of equations) auxiliary variables $\xi_1, \xi_2, \dots, \xi_m$, defined by the equations

$$\xi_i = b_i - \sum_{j=1}^{n} a_{ij}x_j \qquad (i = 1, \dots, m). \tag{55}$$

Examine an auxiliary linear form for the ξ_i:

$$\varphi(\xi) = \xi_1 + \xi_2 + \dots + \xi_n. \tag{56}$$

Let us start minimizing $\varphi(\xi)$ subject to constraints (54) and the assumptions of nonnegativity of the unknowns: $\xi_1 \geq 0$, $\xi_2 \geq 0$, ..., $\xi_m \geq 0$. To solve this problem, we can directly apply the simplex method. Indeed, let us assume that x_j $(j = 1, \dots, n)$ are slack and ξ_i $(i = 1, \dots, m)$ are basic variables. This gives rise to an initial basic feasible solution

$$\xi_i = b_i \quad (i = 1, \dots, m), \qquad x_j = 0 \quad (j = 1, \dots, n).$$

Since ξ_i are nonnegative, $\varphi(\xi) \geq 0$; hence, $\min \varphi(\xi) \geq 0$ too. But $\min \varphi(\xi) = 0$ is attained only when all $\xi_i = 0$. This means that there is a system of nonnegative values of x_j^0 which turn all ξ_i into zero. In other words, the values of x_j^0 $(j = 1, \dots, n)$ satisfy system (54). The reverse is also true: every feasible solution $x_j^0 \geq 0$ $(j = 1, \dots, n)$ of system (54) gives all ξ_i from (55) values equal to zero, i.e., they impart to the form (56) a minimum value $\min \varphi(\xi) = 0$.

Thus, for system (54) to be consistent in the region of feasible solutions (that is, for the polyhedron Q not to be empty), it is necessary and sufficient that $\min \varphi(\xi) = 0$.

Note that when $\min \varphi(\xi) > 0$ it is possible to infer at once that system (54) is inconsistent in the region of nonnegative values x_j $(j = 1, \dots, n)$, i.e., it has no feasible (let alone basic feasible) solution.

Thus, in order to find a feasible solution to an initial linear programming problem, one should minimize the auxiliary form $\varphi(\xi)$. After finding a feasible solution, the same simplex technique can be used to pass from a feasible solution to a *basic* feasible solution and then go on to solve the initial problem (see, e.g., [**6**]).

7

Game Models

7.1. Meteors vs. Pennants (on a soccer theme). The final match for a place in a higher league between the Meteors and the Pennants is drawing to a close. It is a tie. Each team has two spares—a forward and a fullback, but only one substitution is permitted. The Meteors' coach, Smart, knows that his forward can outdo the Pennants' spare fullback but does not measure up to their spare forward, while the Meteors' spare fullback can take care of the Pennants' forward but is not as good as their fullback. Smart puts together the following table:

		"Pennants"	
		B_1	B_2
"Meteors"	A_1	-1	3
	A_1	1	-1

in which the positive numbers correspond to the Meteors' lead, and the negative numbers correspond to the Pennants' lead (in certain arbitrary points). The Meteors' coach starts reasoning like this: "If I move in my forward (let us call it strategy[21] A_1) and Nimble, the Pennants' coach, gets wind of it, he, too, will move in his forward (strategy B_1) and we shall lose, getting -1 (i.e., the Pennants will lead). But if I move in my fullback (strategy A_2), he will move in his fullback (strategy B_2), and we shall lose again, getting -1. But Nimble does not know what I am going to do. Therefore, he will think that I will oppose his forward with my fullback, and we will get a lead of 3."

Smart, who knows something of the theory of games, decides to move in the fullback. How does he arrive at this decision?

By reasoning like this: "Suppose that p is the probability (frequency) that, in situations similar to this, I move in my forward. Then $1-p$ is the probability that in the same kind of situation I move in my fullback. In that case, if Nimble moves in his forward, we shall acquire an average lead equal to

$$v_1(p) = -1 \cdot p + 1(1-p) = 1 - 2p,$$

[21] For a more detailed discussion of the concept of strategy see 7.2.

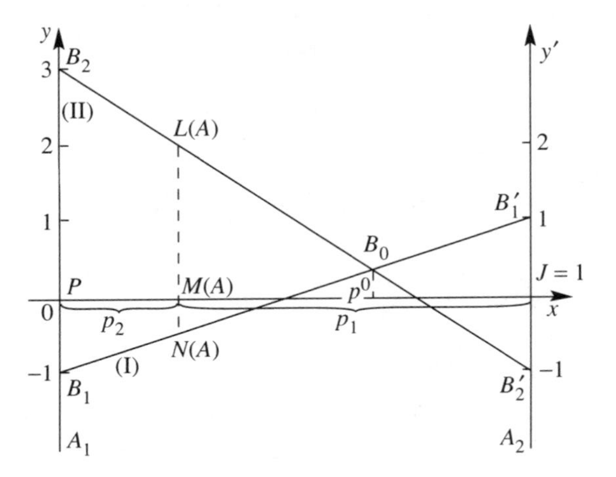

FIGURE 20

and if he moves in his fullback, then we shall get an average lead of

$$v_2(p) = 3p - 1(1 - p) = 4p - 1."$$

Thus, Smart has to find a value of p from the segment $[0, 1]$, such that the least value of two variables, $v_1(p)$ and $v_2(p)$, is the greatest. In other words, he has to find a p that maximizes the minimum of two quantities: $v_1(p)$ and $v_2(p)$. It is evident that as p increases, so does $v_2(p)$, but $v_1(p)$ decreases; and vice versa, as p decreases, $v_1(p)$ increases, but $v_2(p)$ will decrease. Smart decides to represent these situations graphically (Figure 20).

For the x-axis, he chooses the horizontal axis and lets it represent p. He lays off a unit segment $[0, J]$ from point O. He identifies the left end (O) with the strategy A_1, and the right end (J) with A_2. Smart knows that it will not pay for him to adhere throughout to just one of these strategies (a *pure strategy*). Indeed, if he uses A_1 with probability $p(A_1) = 0$ (i.e., not at all) and A_2 with probability $p(A_2) = 1$, then Nimble will always move in a fullback. That is why Smart turns to *mixed strategies* A (for a more detailed discussion see 7.2) using the pure strategies A_1, A_2 with probabilities $p(A_1) = p = p_1$ and $p(A_2) = p_2 = 1 - p$. To each mixed strategy (p_1, p_2) Smart juxtaposes point M_A on the line segment $[0, J]$, defining it so that the distance $|OM(A)| = p_2$, $|M(A)J| = p_1$. Then through the ends of the line segment $[0, J]$ he draws a pair of vertical axes O_y and $J_{y'}$ perpendicular to the axis Ox. He lets the first of these axes represent the payoffs of the pure strategy A_1, intending the second to represent the payoffs under A_2. Should Nimble follow his pure strategy B_1, the Meteors' payoff would be -1 (point B_1); should he choose the strategy B_2, the payoff would be 3 (point B_2). This is so provided that Smart chooses the strategy A_1. If, on the other hand, he chooses the strategy A_2, he will gain either 1 (Nimble using the strategy B_1) or -1 (Nimble using B_2). Smart represents these payoffs on the axis $J_{y'}$ with points B_1' and B_2', respectively. Then he joins with a straight line B_1B_1'(I) the points B_1 and B_1'. The line (I) passes through B_1 with the coordinates $(0, -1)$ and B_1' with the coordinates $(1, 1)$. Hence its equation in the coordinate system Oxy is $y = 2x - 1$. Whatever mixed strategy $A = (p_1, p_2)$ Smart may use, he obtains a payoff corresponding to the point $N(A)$ on the straight line B_1B_1'. Indeed, the ordinate $y(A)$ of $N(A)$ on the line (I) is equal to $y(A) = 2p - 1 = 2(1 - p_1) - 1 - 2p_1 = v_1(p)$.

Then, in a similar way, he draws a straight line B_2B_2' (II) (Nimble's strategy B_2) and marks on it a point, $L(A)$, whose ordinate is the payoff, provided he uses the mixed strategy A and Nimble uses the strategy B_2. In the resulting drawing v_2p is represented by the point $L(A)$. Indeed, the straight line (II) passes through points $B_2(0, 3)$ and $B_2'(0, -1)$ and is determined by the equation $y = -4x + 3$. Given $x = p_2$, the ordinate of $L(A)$ is equal to $y(A) = -4p_2 + 3 = -4(1 - p_1) + 3 = 4p_1 - 1 = v_2(p)$.

Smart is looking for an optimal strategy, $A_0 = (p_1^0, p_2^0)$, that would—assuming that his opponent plays so as to cause him the most harm—maximize his minimum payoff. He sees at once that the line $B_1B_0B_2'$ corresponds to the values of the minima of two variables $v_1(p)$ and $v_2(p)$ for different p from the segment $[0, 1]$ and that the maximum of these minima is reached at the point B_0, i.e., at the intersection of the straight lines (I) and (II).

Smart finds this intersection point and the respective value of p^0, solving the equation

$$v_1(p) = v_2(p),$$

or

$$-1 \cdot p + 1(1 - p) = 3p - 1(1 - p).$$

The solution to this equation is $p^0 = \frac{1}{3}$. Thus, on the average, a forward should be put into play only in one case in three, and a fullback the rest of the time. For that very reason, Smart decides to move in his fullback. Note that such a *maximin* value of $v_1(p)$ and $v_2(p)$ is the *average payoff* for the Meteors' coach (a negative payoff is a loss). In our case the payoff, given $p^0 = \frac{1}{3}$, amounts to

$$v = v_1(\tfrac{1}{3}) = v_2(\tfrac{1}{3}) = \tfrac{1}{3}.$$

And what does Nimble—the Pennants' coach—do? He reasons in a similar way (remember that his payoff is expressed in negative numbers): "Suppose that with probability $q = q_1$ I move in my forward and with probability $1 - q = q_2$ I move in my fullback. If Smart puts into play his forward, our payoff will amount to

$$v_1(q) = -1 \cdot q + 3(1 - q) = -4q + 3 = 4q_2 - 1,$$

and if he puts into play a fullback, then our lead will be

$$v_2(q) = 1 \cdot q - 1(1 - q) = 2q - 1 = -2q_2 + 1."$$

Then Nimble also makes a drawing (Figure 21 on the following page), like Smart's. To every point $T(B)$ of the unit segment $[0, J]$ he juxtaposes his mixed strategy $B = (q_1, q_2)$, so that $|OT(B)| = q_2$, $|T(B)J| = q_1$. He lets

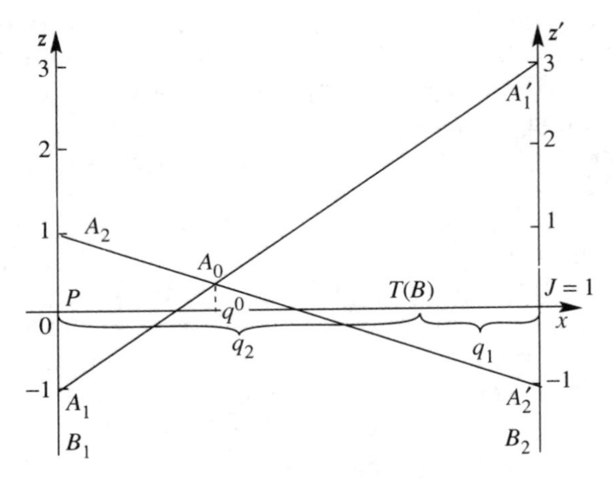

FIGURE 21

the axes Oz and Oz' represent payoffs corresponding to the pure strategies B_1 and B_2, respectively. If he chooses the strategy B_1 and Smart moves in his forward (the strategy A_1), Nimble's payoff will be -1 (point A_1). If, however, Smart moves in his fullback, Nimble's payoff will amount to 1 (point A_2). If only the strategy B_2 is used, there occur on the axis Oz' points A_1' (a payoff of 3) and A_2' (a payoff of -1). Nimble has calculated that in the coordinate system xOz, the ordinates of lines A_1A_1' and A_2A_2' determine $v_1(q)$ and $v_2(q)$, respectively. What Nimble wants is to find such a q from $[0, 1]$ as would minimize the maximum of $v_1(q)$ and $v_2(q)$, since the smaller it is, the better for the Pennants. The positive numbers are the Meteors' payoff, and the negative numbers are the Pennants' payoff. A glance at the drawing shows the Pennants' coach that line $A_2A_0A_1'$ corresponds to the greater of the pair of numbers $v_1(q)$ and $v_2(q)$ for different values of q from line segment $[0, 1]$ and that point A_0 corresponds to the least of these maxima, whereby at an appropriate value of q^0 are reached *minimax* values of $v_1(q)$ and $v_2(q)$. Solving the equation

$$v_1(q) = v_2(q),$$

or

$$-4q + 3 = 2q - 1,$$

he finds that $q^0 = \frac{2}{3}$, i.e., he should, on the average, move in a forward in two cases out of three, and a fullback only in one. The Pennants' average payoff will amount in that case to

$$v(q) = v_1(q) = v_2(q) = \tfrac{1}{3}.$$

Hence, the team loses $\frac{1}{3}$, since for it a win is a negative number and a loss a positive one. It is easy to see that the Meteors gain, on the average, is exactly as much as the Pennants lose, and vice versa, i.e., the sum of the two teams' payoffs is equal to zero. Games in which the total payoff equals zero are called *zero-sum games.*

Thanks to Smart's stratagems, the reader has now been introduced to the methods of decision making based on the mathematical *theory of games.* This acquaintance will be all the more edifying, if bolstered by additional information. And that is what we shall now proceed to supply.

7.2. Matrix games. The mathematical model we have just examined is an example of a *finite, two-person, zero-sum game.* Two-person games are central to the theory of games, and their study makes it possible to grasp the principles of the theory and some of its basic results.

Let us give a more precise description of the concept of a finite, two-person, zero-sum game. The game involves the opposing interests of two players, I and II. It is a set of rules describing the essence of the conflict. The *rules of the game* are the conditions determining each player's moves at every stage of the game, the information about the opponent's behavior, and the payoff for each player. Play of the game consists of a series of moves that have to be selected from among those allowed by the rules. If a player chooses his course of action himself, he is said to make *personal moves.* In chess, for example, all moves are personal. A *chance move* is one selected by a random device (such as the roll of dice). A player's *strategy* is a set of instructions which determine every personal move of the player under every conceivable circumstance during a play. Such strategies are called *pure.*

Now let player I have m pure strategies (moves) $\alpha_1, \dots, \alpha_m$, and let player II have n pure strategies $\beta_1, \dots, \beta_n$. Assume that if the first player chooses an ith strategy α_i and the second player chooses a jth strategy β_j, then the first player wins a_{ij} arbitrary units[22], and the second wins, accordingly, $b_{ij} = -a_{ij}$. It is because $a_{ij} + b_{ij} = 0$ for any i and j that these are called *two-person, zero-sum games*, or *antagonistic games.* In what follows, we shall practically always speak of payoff, since a loss is a negative payoff.

Thus, the set of different outcomes, i.e., payoffs, in this game can be given by the *payoff matrix*

$$A = \begin{pmatrix} a_{11} & a_{12} & \cdots & a_{1n} \\ a_{21} & a_{22} & \cdots & a_{2n} \\ \dots & \dots & \dots & \dots \\ a_{m1} & a_{m2} & \cdots & a_{mn} \end{pmatrix},$$

which indicates a payoff for player I. In other words, the element a_{ij} of A is the first player's payoff if he uses the strategy α_i while the second player uses the strategy β_i. Under these circumstances, player I sets out to maximize his payoff while player II seeks to minimize his (recall that the matrix A determines the first player's gain and the second player's loss).

[22] A gain (loss) does not always have an exact quantitative value. Nevertheless it can practically always be assessed in some way, by applying a certain scale. For example, a win, loss or draw at chess may be assessed at 1, -1, or 0, respectively.

Examine the following game.

$$\begin{array}{c} \text{Player II} \\ \begin{array}{cc} & \begin{array}{cccc} \beta_1 & \beta_2 & \beta_3 & \beta_4 \end{array} \\ \text{Player I} \begin{array}{c} \alpha_1 \\ \alpha_2 \\ \alpha_3 \\ \alpha_4 \\ \alpha_5 \end{array} & \begin{pmatrix} 18 & 3 & 0 & 2 \\ 0 & 3 & 8 & 20 \\ 5 & 4 & 5 & 5 \\ 16 & 4 & 2 & 25 \\ 9 & 3 & 0 & 20 \end{pmatrix} \end{array} \end{array}$$

Let us examine it from the first player's point of view. If he knew what choice player II would make, it would be easy for him to select his best response from the following table:

2nd player's strategy	β_1	β_2	β_3	β_4
1st player's best response	α_1	α_3 or α_4	α_2	α_4
1st player's payoff	18	4	8	25

But since the first player's response depends on the choice the second player will make, and the first player, naturally, does not know what that is going to be, he is not sure what he should do. Hence the first player should try to figure out what specific strategy his opponent will choose upon analyzing the situation. Therefore, player I compiles a similar table of the best choice of strategy by player II:

1st player's strategy	α_1	α_2	α_3	α_4	α_5
2nd player's best response	β_3	β_1	β_2	β_3	β_3
2nd player's payoff	0	0	4	2	0

We can see from the second table that if player I chooses α_1, he is guaranteed 0, just as when he chooses α_2 and α_5. If he chooses α_3, he is guaranteed 4; and if he chooses α_4, he is guaranteed 2. Let us call the least sum that he gets in choosing a strategy the *guaranteed level* of that choice. The greatest of these levels is called the *maximum guaranteed level* of the *best guaranteed result.* In this example, if he chooses α_3, the first player can guarantee for himself a payoff of at least 4, and no other strategy can ensure a payoff of 4.

Adopting the second player's point of view, we infer from the first table that the strategies β_1, β_2, β_3, and β_4 provide, respectively, for the following guaranteed levels: $+18$, $+4$, $+8$, and $+25$. Thus the strategy β_2 guarantees a maximum of 4 (note, once again, that the smaller the number, the better it is for player II, since the guaranteed payoffs amount in this situation to -18, -4, -8, and -25).

Thus, there is every reason to expect player I to apply the strategy α_3 and player II to apply the strategy β_2. We can say that the set of strategies α_3 and β_2 is optimal in the sense that neither player should alter his choice, if the other player does not.

In the general case, a pair of pure strategies α_{i_0} and β_{i_0} is called *optimal* provided that

$$\max_i \left(\min_j a_{ij}\right) = \min_j \left(\max_i \alpha_{ij}\right) = a_{i_0 j_0}$$

(it is clear that $\min \max a_{ij} = \max \min(-a_{ij}) = \max \min b_{ij}$).

Unfortunately, such a pair of pure strategies exists far from always; the situation involving two soccer team coaches that was described earlier is a case in point.

Let us now return to the general case where the game is described by a payoff matrix $A = (a_{ij})$. If player I chooses the strategy α_1, his payoff is sure not to be less than the minimum element in the first row of this matrix, that is, he will win at least $\max_i a_{1j}$. If he chooses α_i, he will win at least $\min_j \min_i a_{ij} = \underline{v}$.

However, player I is free to choose any of his strategies. He may, therefore, choose α_i so as to guarantee for himself the least of the possible payoffs, i.e.

$$\max_j \min_i a_{ij} = \underline{v}.$$

This quantity is known as the *lower value of the game* or a *maximin payoff* (*maximin*). The corresponding strategy (defined by the row that contains $\underline{v}$) is called a *maximin* strategy.

Similarly, payoffs to player II are elements of the matrix A, taken with a minus sign. Thus, once this player applies the strategy β_j, he obtains not less than $\min_i(-a_{ij})$. Naturally enough, he tries to maximize this payoff, resorting to different strategies β_j, i.e., he arrives at

$$\max_j \min_i(-a_{ij}).$$

It is easy to see that for any function $f(x)$

$$\max_x[-f(x)] = -\min_x f(x) \quad \text{and} \quad \min[-f(x)] = -\max_x f(x).$$

Applied to the situation we are investigating, the latter means that

$$\max_j \min_i(-a_{ij}) = \max_j \left[-\max_i a_{ij}\right] = -\min_j \max_i a_{ij}.$$

Hence, player II can choose his strategy so as to obtain necessarily no less than

$$-\min_j \max_i a_{ij}.$$

And, of course, player I will get at most

$$\min_i \max_j a_{ij} = \overline{v}.$$

Hence, even though player I can guarantee himself a payoff not less than

$$\max_i \min_j a_{ij} = \overline{v}\,,$$

player II will no allow the first player's payoff to exceed

$$\overline{v} = \min_j \max_i a_{ij}.$$

The number $\overline{v}$ is called the *upper value of the game* or *minimax*, and the corresponding strategy (defined by the column in which $\overline{v}$ is located) is called the *minimax* strategy.

It may happen (as in the above example) that

$$\max_i \min_j a_{ij} = \min_j \max_i a_{ij} = v\,,$$

i.e., the lower value $\underline{v}$ is equal to the upper value $\overline{v}$. This general value is called the *value of the game* and is equal to a certain element $a_{i_0 j_0}$ of the matrix A. In that case player I is aware that he can win at least v, while player II will try to prevent him from realizing more than v. Therefore, player I will adhere to strategy α_{i_0}. Simultaneously player II can win not less than $-v$, once he adopts strategy β_{j_0}. The element $a_{i_0 j_0} = v$ of the payoff matrix ($a_{32} = 4$ in our example) is simultaneously minimal in row i_0 and maximal in column j_0. Such an element is called a *saddle point* of the matrix A, the game is called a *game with a saddle point*, the strategies α_{i_0} and β_{j_0} are *optimal* for players I and II, respectively, and the pair $(\alpha_{i_0}, \beta_{j_0})$ is called the *solution of the game.* We arrive at the following conclusion.

The solution of a game with a saddle point is stable, i.e., neither player can profit by deviating from his optimal strategy, if the opponent continues to adhere to his.

Thus, if a game with a saddle point is repeated many times over, an optimal strategy guarantees a player a maximum average payoff, always assuming that his opponent plays in the best possible way.

Which is greater—"maximin" or "minimax"?

The soccer motif, with which we started our investigation of games, was associated with the matrix $\left(\begin{smallmatrix} -1 & 3 \\ 1 & -1 \end{smallmatrix}\right)$.

It has no saddlepoint. Indeed,

$$\max_i \min_j a_{ij} = \max_i \left[\min a_{1_j}\,;\, \min a_{2_j}\right] = \max[-1\,;\,-1] = -1 = \underline{v}\,,$$

while

$$\min_j \max_i a_{ij} = \min_j \left[\max a_{i1}\,;\, \max_{i2}\right] = \min[1\,;\,3] = 1 = \overline{v}.$$

Hence $\max_i \min_j a_{ij} \neq \min_j \max_i a_{ij}$.

Let the function $f(x\,;\,y)$ of two variables be determined for each pair of integral values $x = 1, 2, \ldots$ and $y = 1, 2, \ldots$ such that $f(i, j) = a_{ij}$. Let us prove that for any matrix $A = (a_{ij})$ the following inequality is true

$$\max_i \min_j f(x, y) \le \min \max_i f(x, y).$$

By definition of the minimum for any fixed x and y

$$\min_y f(x, y) \leq f(x, y),$$

and by definition of the maximum

$$f(x, y) \leq \max_x f(x, y).$$

Hence,

$$\min_y f(x, y) \leq \max_x f(x, y).$$

But the left-hand side of this inequality does not depend on y; therefore,

$$\min_y f(x, y) \leq \min_y \max_x f(x, y).$$

However, the right-hand side of the last inequality does not depend on x. Therefore,

$$\max_x \min_y f(x, y) \leq \min_y \max_x f(x, y).$$

To complete the proof, it only remains to recall that $f(x, y) = f(i, j) = a_{ij}$ $(i = 1, \dots, n;\ j = 1, \dots, n)$.

As a corollary of what has been proved, we obtain that in a game defined by any matrix A, the lower value $\underline{v} = \max_i \min_j a_{ij}$ does not exceed the upper value $\overline{v} = \min_j \max_i a_{ij}$, i.e., $\underline{v} \leq \overline{v}$. The value v of the game lies between $\underline{v}$ and $\overline{v}$.

It can also be shown (we shall not do it here) that the necessary and sufficient condition for realizing the equality

$$\max_i \min_j a_{ij} = \min_j \max_i a_{ij}$$

is that the matrix A have a saddle point. For example, the matrix

$$\begin{pmatrix} 20 & 9 & 30 \\ 31 & 0 & 4 \end{pmatrix}$$

has a saddlepoint $a_{12} = 9$—the smallest element in the first row and the greatest in the second column.

There are two saddlepoints, $a_{11} = 7$ and $a_{13} = 7$, in the matrix

$$\begin{array}{c} \\ \alpha_1 \\ \alpha_2 \end{array} \begin{array}{c} \begin{array}{ccc} \beta_1 & \beta_2 & \beta_3 \end{array} \\ \begin{pmatrix} 7 & 10 & 7 \\ 11 & 24 & 15 \end{pmatrix} \end{array}.$$

Playing with a payoff matrix like that, player I has to follow the strategy α_1, while player II can choose either β_1 or β_3.

If a game has no saddlepoint, then neither player has a unique most reliable pure strategy. At the same time, a player cannot divine the opponent's strategy, if the latter, rather than adhere to a unique pure strategy, starts using a set of several pure strategies, alternating them randomly with a definite frequency relation. Such a set is called a mixed strategy. Thus, a *mixed strategy* is a vector $x = (x_1, x_2, \dots, x_n)$, whose component x_i $(i = 1, 2, \dots, n)$

is the probability that a player chooses an ith strategy α_i out of the set $\alpha_1, \alpha_2, \dots, \alpha_n$ of his pure strategies.

There was no saddlepoint in the game between the coaches Smart and Nimble. That was why both resorted to mixed strategies.

Let us dwell a little on the concept of mixed strategy. Let $x_1, x_2, \dots, x_m$ be the probabilities that the first player chooses the strategies $\alpha_1, \alpha_2, \dots, \alpha_m$, and let $y_1, y_2, \dots, y_m$ be the probabilities that the second player chooses the strategies $\beta_1, \beta_2, \dots, \beta_m$. It is obvious that

$$x_i \geq 0 \quad \text{for } i = 1, \dots, m \quad \text{and} \quad \sum_{i=1}^{m} x_i = 1,$$

$$y_j \geq 0 \quad \text{for } j = 1, \dots, n \quad \text{and} \quad \sum_{y=1}^{n} y_j = 1.$$

In that case, the average payoff for the first player (and loss for the second player) is determined by the formula (it determines the expected payoff):

$$v(x, y) = (x_1, \dots, x_m) \begin{pmatrix} a_{11} & a_{12} & \cdots & a_{1n} \\ \cdots & \cdots & \cdots & \cdots \\ a_{m1} & a_{m2} & \cdots & a_{mn} \end{pmatrix} \begin{pmatrix} y_1 \\ \vdots \\ y_n \end{pmatrix} = \sum_{j=1}^{n} \sum_{i=1}^{m} a_{ij} x_i y_j.$$

Note that written here is the matrix product: the row matrix $x = (x_1, \dots, x_m)$ is multiplied by the matrix A, and the product is then multiplied by the column matrix y comprised of $y_1, \dots, y_n$.

The creator of the modern theory of games, J. von Neumann (1903–1957), provided the proof of the cornerstone of the theory, the *minimax theorem* (its proof can be found, for example, in **[17]**).

Von Neumann's minimax theorem states that there is at least one pair of vectors, x_0 and y_0, satisfying the constraints mentioned above and such that

$$\max_x \min_y v(x, y) = \min_y \max_x v(x, y) = v(x_0, y_0) \equiv v,$$

where v is the value of the game.

In some games (chess, checkers, tic-tac-toe, etc.) each player makes his move, knowing the results of all previous moves made by both players. According to the general theory of games, every game of this type, known as the *perfect information game*, has a saddle point and thus has a solution in pure strategies, resulting, on the average, in a payoff equal to the value of the game v.

However, more frequently matrices have no saddle points. In such games in order to achieve a maximum average payoff, the players have to stick to their maximin and minimax strategies. Furthermore, player I guarantees for himself a payoff equal to the lower value of the game $\underline{v}$. If, however, he resorts to a suitable mixed strategy, he will be able to increase his payoff to the value of the game $\overline{v}$. In practice, when mixed strategies are used, some system of random search (for example, a 0-to-1 random number device or the roll of dice), capable of indicating the occurrence of each pure strategy with the probability pertaining to it (as a component of the mixed strategy), is realized before each play of the game. The pure strategy selected in this way before each successive play is the one used in that play.

In the case of mixed strategies, for each play with a finite $m \times n$ matrix A there is (and can be found) a pair (x_0, y_0) of optimal strategies. That is just what the fundamental minimax theorem states. Optimal strategies have stability in the sense that if one of the players confines himself to his optimal strategy, it does not pay for the other player to deviate from his own optimal strategy.

Suppose that we have found the solution for a game—a pair of optimal strategies $x_0 = (x_1^0, \dots, x_m^0)$, $y_0 = (y_1^0, \dots, y_n^0)$. Those of the pure strategies α_i (respectively, β_j) to which correspond zero values of probabilities $x_i^0 = 0$ $(y_j^0 = 0)$, are called *passive*; the rest are called *active strategies.*

Let us make certain that, if one of the players adheres to his optimal strategy x^0, then his payoff does not change. It is equivalent to the value of the game v, regardless of what the second player may do, if he uses only his active pure strategies.

Indeed let the active pure strategies be $\alpha_1, \dots, \alpha_s$ (for player I) and $\beta_1, \dots, \beta_r$ (for player II); and let the rest of the strategies be passive. Hence,

$$x_0 = (x_1^0, \dots, x_s^0, 0, \dots, 0), \qquad x_i^0 > 0 \ (i = 1, \dots, s),$$
$$y_0 = (y_1^0, \dots, y_r^0, 0, \dots, 0), \qquad y_j > 0 \ (j = 1, \dots, r).$$

Besides, $x_1^0 + \cdots + x_s^0 = 1$, $y_1^0 + \cdots + y_r^0 = 1$. Suppose that player I uses his optimal strategy x_0, and player II uses any mixed strategies comprised only of the active pure strategies $\beta_1, \dots, \beta_r$. If player I uses his optimal strategy x_0 and player II uses his active pure strategy β_j, then player I can obtain a payoff v_j which, as follows from the fundamental theorem, can even exceed the value of the game: $v_j \geq v$. Nevertheless, in fact only an exact equality is possible: $v_j = v$. After all, in the mixed strategy y_0, the pure strategies $\beta_1, \dots, \beta_r$ are used with probabilities $y_1^0, \dots, y_r^0$, respectively. Therefore, payoff expectation (i.e., the average payoff) will amount to

$$v_1 y_1^0 + v_2 y_2^0 + \cdots + v_r y_r^0.$$

Simultaneously, this payoff must be equal to the value of the game

$$v_1 y_1^0 + \cdots + v_r y_r^0 = v, \qquad y_1^0 + \cdots + y_r^0 = 1;$$

whence, it follows that no v_j can exceed v. Otherwise the total will exceed v.

The skeptical reader may ask if the introduction of a mixed strategy is of any fundamental significance. What does it mean to choose a mixed strategy, and would anybody choose it in fact? As for the role played by mixed strategy, that largely depends on one's subjective perception of probability. And as concerns the choice of a "pure" strategy by means of a mixed one, this method is equivalent to staging a probabilistic experiment. Before making the substitution, the Pennants' coach Nimble should first set up a trial with outcome probabilities $\frac{1}{3}$ and $\frac{2}{3}$. He should, for example, roll a die. If the number that falls out is divisible by three—i.e., 6 or 3—he should move in his forward, but the rest of the time it should be the fullback. Of course, this suggestion may seem naive, but in fact, in that situation this method yields optimal behavior. Unfortunately, strategists are too often evaluated by the outcomes, rather than strategic expediency, of their choices in a general risk situation.

7.3. Problem of the final spurt. Let us try to generalize the situation described above and to examine competitive games in which at least one of the opponents has an infinite set of pure strategies. Such games are called *infinite antagonistic games.* Their general theory is much more complex than the theory of finite antagonistic (matrix) games. Therefore, we shall limit ourselves to just one example.

First let us introduce some concepts that we are going to need.

Assume that each player has an infinite set (continuum) of pure strategies. The first player's pure strategies are represented by points of a unit segment $0 \le \xi \le 1$, and the strategies of player II, by points of a unit segment $0 \le \eta \le 1$. In other words, the players' pure strategies are points (numbers) contained between zero and one. The part of the payoff matrix (a_{ij}) in this case is played by a function, $V(\xi, \eta)$, of two arguments, ξ and η. It is called the *payoff function* or *core of a game.* The geometric image of the payoff function is a surface located above a unit square $K = 0 \le \xi \le 1$, $0 \le \eta \le 1$ (Figure 22). When the first player chooses strategy ξ_0 and the second player chooses strategy η_0, the payoff of the first player becomes equal to $V(\xi_0, \eta_0)$—the z-coordinate z_0 of a surface point with the abscissa ξ_0 and ordinate η_0.

Examination of a game with a payoff function is similar to that of a finite game. First we find for each fixed value of ξ from the segment $[0, 1]$ the minimum of $V(\xi, \eta)$ as a function of η: $\min_{0\le\eta\le1} V(\xi, \eta)$. Then we find

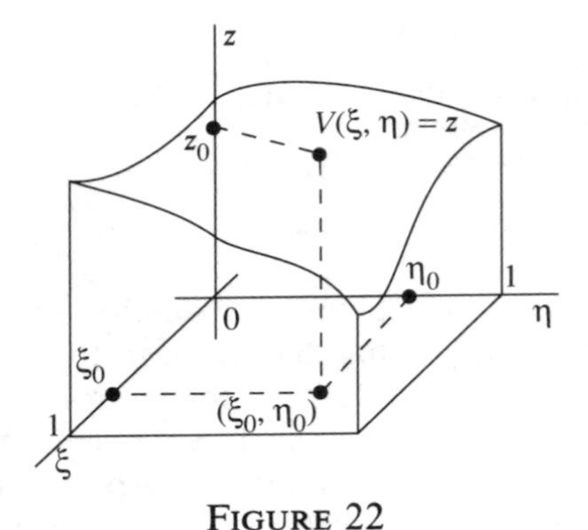

FIGURE 22

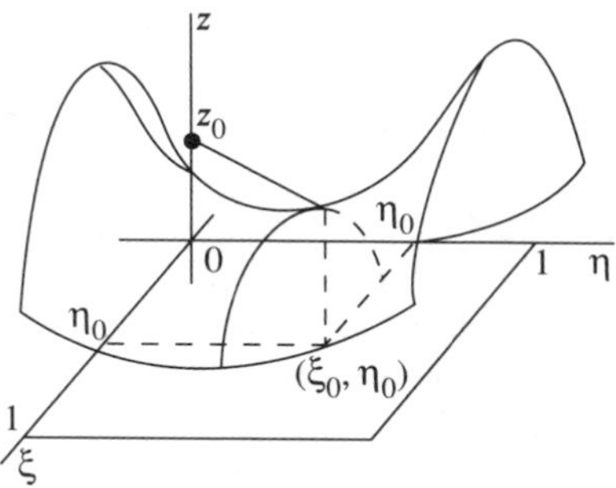

FIGURE 23

the maximum of the values obtained relative to ξ:

$$\max_{0\le\xi\le1}\min_{0\le\eta\le1} V(\xi, \eta) = \underline{v}.$$

This gives the lower value of the game (maximin) $\underline{v}$. Then we find the upper value of the game (minimax):

$$\overline{v} = \min_{0\le\eta\le1}\max_{0\le\xi\le1} V(\xi, \eta).$$

We already know (see 7.2) that the inequality $\underline{v} \le \overline{v}$ is always satisfied. If the upper value of a game is equal to its lower value ($\underline{v} = \overline{v}$), it means that there is a pair of numbers (ξ_0, η_0) for which

$$\min_{\eta}\max_{\xi} V(\xi, \eta) = \max_{\xi}\min_{\eta}(v(\xi, \eta) = V(\xi_0, \eta_0).$$

In this case, the game has a solution in pure strategies: the first player has to keep to his strategy $\xi = \xi_0$ and the second, to his strategy $\eta = \eta_0$. These are optimal strategies. The first player gets $V(\xi_0, \eta_0)$, and the second player gets $-V(\xi_0, \eta_0)$. The relevant point with the coordinates $(\xi_0, \eta_0, z_0 = V(\xi_0, \eta_0))$ is called a *saddlepoint.* It is the point where the minimum in η and maximum in ξ are reached simultaneously (see Figure 23). It makes no sense for either player to deviate from his optimal strategy: that can only result in diminishing his payoff (if the opponent continues in his optimal strategy).

When the upper value of a game exceeds the lower value, that game has no solution in pure strategies. The solution should only be sought in mixed strategies. It means that in the situation with which we are dealing, the strategies ξ and η are regarded as random variables and are assigned by their functions of probability distribution $P_1(\xi)$ and $P_2(\eta)$. The value of function $P_1(\xi_1)$ at a fixed $\xi = \xi_1$ gives the probability that the number ξ, selected at random on the segment $[0, 1]$, will be less than ξ_1. Function $P_2(\eta)$ has a similar meaning: its value $P_2(\eta)$, given η_1, is equal to the probability that for the number η, selected at random on the segment $[0, 1]$, an inequality $\eta < \eta_1$ is realized.

The theory of infinite games proves that if the payoff function $V(\xi, \eta)$ is continuous, there always are optimal strategies (i.e., there always is a solution).

Let us now take up a specific problem. Two speed skaters are competing for a 10,000-meter stayer distance. Each of the skaters realizes that he can risk not more than one spurt. Naturally, each of them notices his rivals' spurt.

Let us introduce function $P_1(\xi)$ determined on a unit segment $0 \leq \xi \leq 1$. Suppose that at each ξ the value of $P_1(\xi)$ determines the probability of success—of a skater winning the race—provided that the skater has spurted not farther away from the start than ξ. Here we have "normalized" the distance, assuming its length to be 1 ($\xi = 0$ corresponds to the start and $\xi = 1$ to the finish). Thus, $P_1(\xi)$ is the probability distribution function of the first skater's success (victory). It is natural to suppose that the success function $P_1(\xi)$ is continuous for all values of ξ out of $[0, 1]$, increasing steadily from $P_1(0) = 0$ to $P_1(1) = 1$. We shall require that similar conditions be satisfied by the function $P_2(\eta)$ which represents the probability distribution of the other skater's success at the same distance.

Let us agree (which is also a kind of normalization) that if player I outdistances player II, then his payoff (success) is $+1$; if they finish simultaneously, it is 0; if player II finishes first, it is -1 for the first player.

Let us now put together a player's payoff function $V(\xi, \eta)$. If $\xi < \eta$, i.e., if player (skater) I spurts earlier than player II, having covered the distance ξ, then the probability that he breaks away and wins is $P_1(\xi)$. It is clear that if player II does not fall behind player I, then after spurting, player II will get ahead and win with probability $P_2(\eta)$, where $\eta = \xi$.

Similar arguments can be applied to player II. If he attempts to get ahead after covering a distance $\eta < \xi$, he will succeed with probability $P_2(\eta)$ or fail with probability $1 - P_2(\eta)$.

Thus, the first player's payoff in arbitrary points amount to

$$\begin{aligned} L(\xi, \eta) &= 1 \cdot P_1(\xi) + (-1)[1 - P_1(\xi)], \quad \text{if } \xi < \eta; \\ M(\xi, \eta) &= (-1)P_2(\eta) + 1 \cdot [1 - P_2(\eta)], \quad \text{if } \xi > \eta; \\ \Phi(\xi) &= 1 \cdot P_1(\xi)[1 - P_2(\xi)] \\ &\quad + (-1)[1 - P_1(\xi)]P_2(\xi), \quad \text{if } \xi = \eta. \end{aligned}$$

Games in which the second player's payoff function (core of the game) is determined by the relations

$$V(\xi, \eta) = \begin{cases} L(\xi, \eta) & \text{for } \xi < \eta, \\ \Phi(\xi) & \text{for } \xi = \eta, \\ M(\xi, \eta) & \text{for } \xi > \eta, \end{cases}$$

where the functions $L(\xi, \eta)$ and $M(\xi, \eta)$ are determined and continuous in a square $0 \leq \xi \leq 1$, $0 \leq \eta \leq 1$, and the function $\Phi(\xi)$ is continuous on the unit segment $0 \leq \xi \leq 1$—such games are called *time-selection* games. Time-selection games occur in many areas of men's activity. They do not always have optimal strategies.

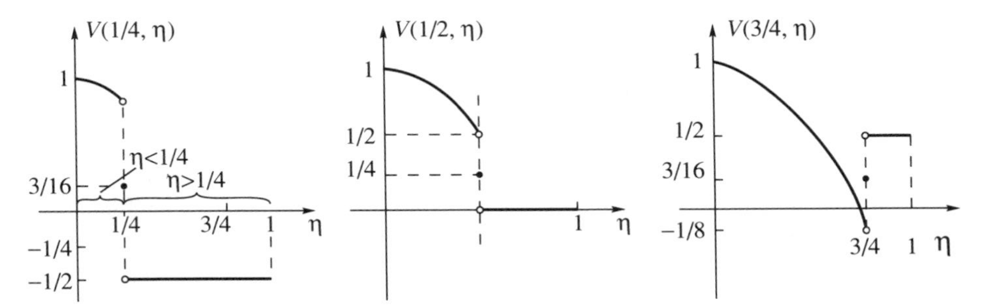

FIGURE 24

In our specific case the payoff function, after simplifications, is

$$V(\xi, \eta) = \begin{cases} 2P_1(\xi) - 1 & \text{for } \xi < \eta, \\ P_1(\xi) - P_2(\xi) & \text{for } \xi = \eta, \\ 1 - 2P_2(\eta) & \text{for } \xi > \eta. \end{cases}$$

Considering $V(\xi, \eta)$ as a function of η, given fixed values of ξ, suggests that player I will have an optimal pure strategy where the constant part of the payoff, i.e., $2P_1(\xi) - 1$ (after all, ξ is fixed) is equal to the least value of the expression $1 - 2P_2(\eta)$ for $\eta < \xi$. This requires that the following equality be realized:

$$2P_1(\xi) - 1 = 1 - 2P_2(\xi).$$

Since the right side of this equality decreases continuously from $+1$ to -1 when ξ belongs to the segment $[0, 1]$, while the left side increases continuously from -1 to $+1$, there exists at least one solution ξ_0 of this equation. Note that if both functions $P_1(\xi)$ and $P_2(\eta)$ are strictly monotonic, then ξ_0 is a unique solution.

Let us assume, for example, that in our problem $P_1(\xi) = \xi$ and $P_2(\eta) = \eta^2$. Then the payoff is

$$V(\xi, \eta) = \begin{cases} 2\xi - 1 & \text{for } \xi < \eta, \\ \xi - \eta^2 & \text{for } \xi = \eta, \\ 1 - 2\eta^2 & \text{for } \xi > \eta. \end{cases}$$

Given different fixed values of ξ, we can represent the lines of intersection of a surface $V(\xi, \eta)$ by planes $\xi = \text{const}$ (see Figure 24).

So, given $\xi = \frac{1}{4}$ (i.e., in section by the plane $\xi = \frac{1}{4}$), we obtain

$$V(\tfrac{1}{4}; \eta) = \begin{cases} 2\frac{1}{4} - 1 = -\frac{1}{2} & \text{for } \eta > \frac{1}{4} \text{ straight line, parallel to the axis } O\eta; \\ \frac{1}{4} - \frac{1}{16} = \frac{3}{16} & \text{for } \eta = \frac{1}{4} \text{ isolated point;} \\ 1 - 2\eta^2 & \text{for } \eta < \frac{1}{4} \text{ arc of parabola.} \end{cases}$$

Let us solve in our problem of the speed skaters the equation

$$2P_1(\xi) - 1 = 1 - 2P_2(\xi),$$

where $P_1(\xi) = \xi$ while $P_2(\xi) = \xi^2$. We obtain

$$2\xi - 1 = 1 - 2\xi^2$$

or

$$2\xi^2 + 2\xi - 2 = 0;$$

whence, we find $\xi_0 = (-1 + \sqrt{5})/2$ (the other root of the quadratic equation will not do since it does not belong to the segment $[0, 1]$).

$$\frac{\sqrt{5} - 1}{2} \cdot 10{,}000 \approx 6{,}180 \text{ meters},$$

and his expected payoff will amount to $2P(\xi_0) - 1 = \sqrt{5} - 2 = 0.21$.

So, subject to our selected probability distribution functions $P_1(\xi)$ and $P_2\eta$, the first skater has more chance than the second of winning the race.

In the general case, this problem can be solved given different probability distribution functions of success $P_1(\xi)$ and $P_2(\eta)$ which can be found empirically or by questioning coaches and skaters.

7.4. Games against nature. In problems concerned with the theory of games the players make decisions under uncertainty in the sense that one does not know how the other will respond. Therefore, the opponent is always assumed to be clever and malicious, so to speak. However, very often the uncertainty is not due to the opponent's deliberate opposition, but rather to our insufficient knowledge of the circumstances in which we have to make decisions. In all such cases, the conditions of the game and the payoff do not depend on a consciously antagonistic opponent, but on objective reality which is commonly described as *nature.* Relevant situations are called *games against nature.* The theory of games against nature is also known as the *statistical decision theory,* in which nature is regarded as a disinterested entity whose behavior is unknown and contains no element of conscious opposition.

Let us have m pure strategies $\alpha_1, \dots, \alpha_m$, while nature may be in one of the n states $\beta_1, \dots, \beta_n$. Then our payoff, just as before, can be assigned by a payoff matrix A with the elements a_{ij}, where $i = 1, \dots, m$ and $j = 1, \dots, n$.

At first sight, it may seem that this is simpler than a game problem, since no opposition is offered. Indeed, in a game against nature the decision maker is likely to get more payoff than in a game against a regular antagonist. But then, it is harder for him to make a valid decision that will yield a sizable payoff. As a matter of fact, in a conflict situation the assumption of the players' antagonism removes, in a sense, uncertainty concerning the opponent's possible reasonable actions. But in a game against nature, uncertainty looms large, since one cannot expect nature to behave "reasonably".

7.5. How to wax skis. Let us examine the following problem. As is common before a competition, the coaches of a skiing team are discussing ski wax. They know that during the race the weather (nature) may be in any of three states, β_1, β_2, or β_3, but they have four kinds of ski wax, α_1, α_2, α_3, and α_4. Matrix A of the probability that the competitors will successfully cover the route is known:

$$\begin{array}{c} \\ \alpha_1 \\ \alpha_2 \\ \alpha_3 \\ \alpha_4 \end{array} \begin{array}{c} \begin{array}{ccc} \beta_1 & \beta_2 & \beta_3 \end{array} \\ \begin{pmatrix} 0.20 & 0.30 & 0.15 \\ 0.75 & 0.20 & 0.35 \\ 0.25 & 0.80 & 0.25 \\ 0.85 & 0.05 & 0.45 \end{pmatrix} \end{array} = A.$$

In other words, it is known that if they use the ski wax α_3 and nature is in the state β_2 then with probability 0.8 this ski wax will enable a skier to cover the distance at full speed. The other elements of the matrix A have a similar meaning. So, which ski wax does one select?

It would be simple for the coaches to solve the problem, if they knew the probabilities p_1, p_2, p_3 of the weather being in any one of the three possible states. Indeed, it is enough to find the expected payoff for each wax and choose the alternative with the greatest average expected payoff. If in this example the probability of the first of nature p_1 were equal to 0.3, of the second state $p_2 = 0.5$, and of the third state $p_3 = 0.2$, then the expected average payoffs would amount, respectively, to

$$m_{\alpha_1} = 0.3 \cdot 0.2 + 0.5 \cdot 0.3 + 0.2 \cdot 0.15 = 0.24,$$
$$m_{\alpha_2} = 0.3 \cdot 0.75 + 0.5 \cdot 0.2 + 0.2 \cdot 0.35 = 0.395,$$
$$m_{\alpha_3} = 0.3 \cdot 0.25 + 0.5 \cdot 0.8 + 0.2 \cdot 0.25 = 0.525,$$
$$m_{\alpha_4} = 0.3 \cdot 0.85 + 0.5 \cdot 0.05 + 0.2 \cdot 0.25 = 0.33.$$

Thus, the choice should be made in favor of the third wax alternative, and the expected payoff in that case would be the largest of all four, namely 0.525.

Unfortunately, the coaches did not know the probabilities concerning the states of the weather (they did not trust the weather forecasts), and so they had to look for the answer in a different way. Now, we must note that the head coach was a rationalist, while one of his assistants was a pessimist and the other an optimist. The pessimist believed that they should choose the alternative that would ensure the greatest minimum payoff. He suggested writing out all the minimum elements in the rows, namely, 0.15, 0.20, 0.25, 0.05, and selecting the largest of them. It was 0.25, and the corresponding strategy, proposed by the pessimist, was α_3. This choice is prescribed by the *Wald maximin test.*

By this test, the strategy α is optimal if its minimum payoff is maximal: $\max_i \min_j a_{ij}$. Thus, this test is oriented at the worst kind of weather and is decidedly pessimistic.

Then the head coach asked the optimist to give his opinion. Naturally the optimist, who always hoped for the best, suggested using the fourth wax alternative which, given favorable conditions, would yield the greatest payoff.

Having heard his assistants, the head coach (a longheaded fellow) said they should apply the *Hurwitz pessimism/optimism criterion*; namely, that they first choose a number $0 \le \lambda \le 1$, and then find the row for which was attained

$$\max_i \left(\lambda \min_j a_{ij} + (1 - \lambda) \max_j a_{ij} \right).$$

It is easy to see that if $\lambda = 1$, it is the Wald pessimism test; and if $\lambda = 0$, it turns into the "extreme optimism" test.

If the value of λ is intermediate, so is the test—it is neither extremely pessimistic nor extremely optimistic. The head coach suggested making λ equal to 0.6, to allow for a possible worsening of the weather. After that, they found the values of h_i $(i = 1, \dots, 4)$ for each row

$$h_1 = 0.6 \cdot 0.15 + 0.4 \cdot 0.3 = 0.21,$$
$$h_2 = 0.6 \cdot 0.2 + 0.4 \cdot 0.75 = 0.42,$$
$$h_3 = 0.6 \cdot 0.25 + 0.4 \cdot 0.8 = 0.47,$$
$$h_4 = 0.6 \cdot 0.05 + 0.4 \cdot 0.85 = 0.37.$$

The greatest was $h_3 = 0.47$, so everybody was inclined to use the third wax alternative (the strategy α_3). From the standpoint of common sense, too, such a decision was justified, since in strategic decision making one should be neither extremely optimistic nor extremely pessimistic.

It seemed that the question was settled. But taking the floor at the last minute was a member of the coaches' council, something of a "stormy petrel", who had his own ideas about everything. This time too, he declared that they did not need to study the elements of the success matrix to make a decision. Why? "First of all," he said, "the strategy α_1 should be excluded from consideration right away because it is demonstrably worse than α_3. Indeed, whatever the state β_j of nature, the probabilities of success $a_{11} = 0.20$, $a_{12} = 0.30$, and $a_{13} = 0.15$ are less than the corresponding probabilities $a_{31} = 0.25$, $a_{32} = 0.80$, and $a_{33} = 0.25$. In other words, the first row of the matrix A should be discarded and we should work with a success matrix of a smaller dimension:

$$\tilde{A} = \begin{matrix} & \beta_1 & \beta_2 & \beta_3 & \\ \Big\{ & 0.75 & 0.20 & 0.35 & \Big\} \alpha_2 \\ & 0.25 & 0.80 & 0.25 & \alpha_3 \\ & 0.85 & 0.05 & 0.45 & \alpha_4 \end{matrix}.$$

"Second and most important, there are generally no valid reasons to use the success matrix. For example, the success $a_{41} = 0.85$, given a strategy α_4 and a state of nature β_1, is greater than given the strategy α_3 and the state β_2. However, choosing α_4 over α_3 may be better merely because the state β_1 is more favorable than β_2. Thus we compare our decisions in different situations β_j. In fact, however, we should argue in such a way as to be able to take into account the benefits we can derive from each state of the weather, subject to different choices of α_i. Namely, if we knew in advance what the true state of the weather β_j was going to be, then the choice of α_i would be prefixed. We would simply choose in the column for β_j the maximum element, $\max_i a_{ij}$. Let us denote it by $c_j = \max_i a_{ij}$. Now let us take the difference $c_j - a_{ij} = r_{ij}$. This difference shows how the payoff a_{ij} that we would get if we chose the strategy α_i (being ignorant of the state of the weather β_j) deviates from the maximum possible payoff c_j that we get when we are informed in good time that nature is in the state β_j. Therefore, we will naturally consider the difference r_{ij} to be our "risk" in using the strategy α_i while the weather is in the state β_j. And it is much more sensible to look to the risk matrix $R = (r_{ij})$ rather than to the success matrix A. For this purpose let us find

$$c_1 = \max_i a_{i1} = 0.85, \qquad c_2 = \max_i a_{i2} = 0.80,$$
$$c_3 = \max_i a_{i3} = 0.45$$

and compute the differences r_{ij}. We obtain the risk matrix

$$R = \begin{array}{c} \begin{array}{ccc} \beta_1 & \beta_2 & \beta_3 \end{array} \\ \left\{ \begin{array}{ccc} 0.10 & 0.60 & 0.10 \\ 0.60 & 0.00 & 0.20 \\ 0.00 & 0.75 & 0.00 \end{array} \right\} \end{array} \begin{array}{c} \alpha_2 \\ \alpha_3 \\ \alpha_4 \end{array}.$$

Compare it with the matrix $\widetilde{A}$. In this matrix payoffs a_{31} and a_{33} are identical, being equal to 0.25. However, they are far from being equivalent, since if the strategy α_3 is chosen, given the state of nature β_1, the risk $r_{31} = 0.60$ is three times the risk $r_{33} = 0.20$ incurred with the same strategy α_3, given the state β_3. "For these reasons," he continued, "whenever the state of nature is not defined in advance, we should strive above all to diminish the risk. In other words, we should choose a strategy that provides for the least risk in the worst of circumstances. And for this it is necessary to calculate $\min_i \max_j r_{ij}$.

"In our conditions $\max_j r_{2j} = 0.60$, $\max_j r_{3j} = 0.60$, $\max_j r_{4j} = 0.75$, and $\min_i \max_j r_{ij} = \min_i\{0.60; 0.60, 0.75\} = 0.60$. Therefore, we should choose the strategy α_2 or α_3 that guarantees the least risk, regardless of the weather (in the least favorable circumstances)."

"Excuse me," said the head coach, interrupting the speaker, "but your suggestion is little better than the pessimist's. The only difference is that by your criterion the worst that can happen is a maximum risk, rather than a minimum payoff. But this is the *Savage minimax criterion*, well known in decision theory. This criterion is as pessimistic as the Wald maximin criterion. So let us keep the golden mean and finally choose the strategy α_3. As for simplifying the matrix A by discarding the strategies that are obviously worse than others, this should have our full support." That ended the discussion on how to wax the skis.

7.6. The iron game. Many peoples have epics describing mighty heroes, capable of lifting and carrying objects of enormous weight. Power exercises and popular matches in lifting and throwing heavy objects go back to antiquity. They have provided the source of the modern competitive sport of *weight lifting.*

European weight-lifting championships have been held annually since 1896, and world championships since 1898. In 1920, the International Weight-Lifting Federation (FHI) was formed and weight lifting became an Olympic sport.

The weight used in competitions is the barbell, a steel bar to which disk weights are attached.

Competitions in weight lifting are conducted following the Olympic system, which consists of two successive two-hand lifts—the snatch and the clean and jerk.

In the snatch, the barbell is lifted from the floor to arm's length overhead in a single movement.

The clean and jerk is a two-part lift. After lifting the barbell to his shoulders, the lifter jerks it overhead to arm's length.

Since a lifter's performance largely depends on his own weight, competitions are held separately in ten weight categories (regardless of age).

Each contestant is entitled to three attempts in the snatch and three in the clean and jerk. The lifter notifies the judges of the weight with which he starts the match.

Nevertheless, he can make his first attempt with a weight either greater or smaller than that declared. After a successful first attempt, a weight of at least 5 kg (11 lb) is added to the barbell. After a successful second attempt and before the third, the weight is increased by at least 2.5 kg (5.5 lb). If all three attempts have been successful, the lifter then has the opportunity to set a record in a fourth attempt. Furthermore, the barbell must be at least 0.5 kg (1.1 lb) heaver than the record weight. The result of the fourth attempt is not credited to competition results.

In individual matches the champion is determined by the sum total of both lifts. Should several athletes lift an identical weight, the champion is the one whose own weight before the competition was less than that of any of the others. If more than one of the lifters had the same weight, then they are weighed also after the competition. Those whose weights coincide yet again, share the same place.

At international and national competitions the judges' panel consists of a judge, two assistant judges, a secretary, and a time-keeper (a maximum of three minutes is allowed between a lifter's being called to the platform and completing a lift). All disputes are settled by a three- or five-man jury.

Let us now examine weight-lifting competitions in terms of game theory.

Suppose that taking part in a match are two main contenders, I and II. After drawing lots, lifter I is to come to the platform each time before his chief rival II. At a competition, a lifter can either attempt to lift the weight or he can postpone his attempt until the weight is increased and, if he likes, until it is further increased. For simplicity's sake, we shall assume that each lifter is entitled to just one attempt or postponement for any one weight. If

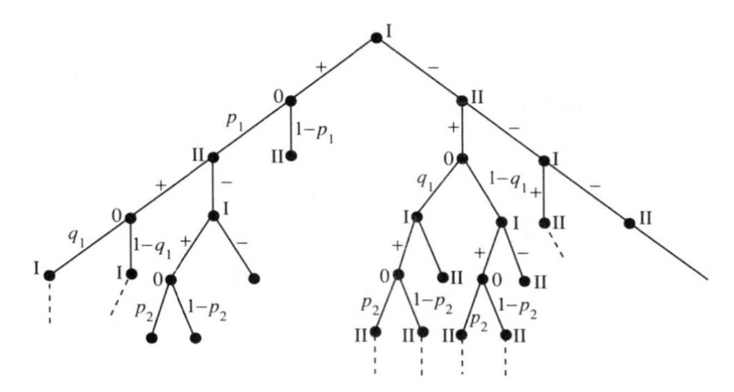

FIGURE 25

it were three attempts, there would merely be more situations to consider, but the model below would not be substantially affected. Incidentally, this is not a very difficult case to analyze using a computer.

Let us assume that the initial weight is 200 kg (440 lb); the probability of it being lifted by I is p_1, and that of it being lifted by II is q_1. The probabilities regarding the next weight, 202.5 kg (445.5 lb), are p_2 and q_2, respectively, and so on. In this case the game can be represented as a tree diagram (graph) shown in Figure 25.

If a vertex is marked with the figure I, it means that in this situation the first player moves (that is, either makes or skips an attempt); if it is marked with the figure II, the second player moves; and if it is marked with 0, then a "random mechanism" is at play, such that with probability p_k or q_k the kth weight ($200\text{ kg}/440\text{ lb}+k\cdot 2.5\text{ kg}/5.5\text{ lb}$) is lifted, and with probability $1-p_k$ (or $1-q_k$) it is not lifted. The plus sign at the arc implies that the respective lifter (this arc emerges from the vertex marked with an appropriate number) has decided to make the attempt, and the minus sign shows that he waits for a heavier weight. Obviously, the construction of the graph can be continued. It must, of course, be remembered that the probabilities p_k and q_k abruptly decrease as the weight increases.

Games plotted as a tree diagram are called *extensive* or *positional* games. This name is due to the fact that at any point in the game each player is informed about his own and his opponent's moves and knows at which vertex of the graph he is.

Pure strategy in a positional game is a set of instructions such that to each vertex of the graph corresponds one definite action by the player whose turn it is to move. The noted mathematician Zermelo proved that for any positional game (including that with random moves, as in our example) there is an *optimal pure strategy*, or a best strategy which indicates what each player has to do in any position (situation) that shapes up in the course of play.

We shall not consider what the optimal strategy may be in the above example; finding it involves an exhaustive search of a great many alternatives and cannot be done without a computer.

We shall merely note that if the probabilities p_k, q_k ($k = 1, 2, \ldots$) are

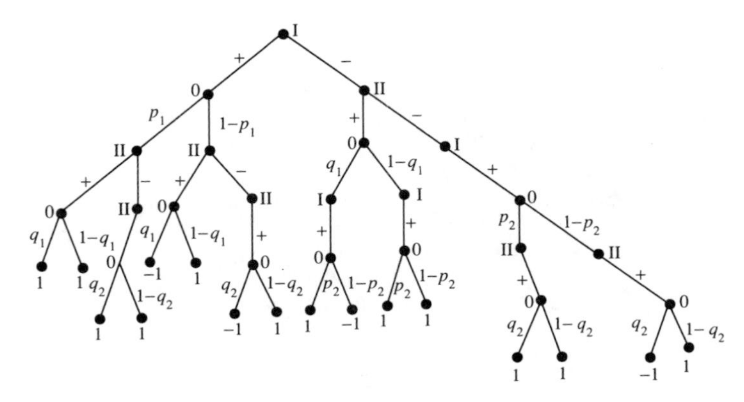

FIGURE 26

known, finding an optimal strategy in the foregoing example is feasible. It seems to us that it is very important psychologically for the lifter and his coach to be aware of what their optimal strategic behavior should be, for it safeguards them at least from a tactical loss, though of course it does not, on the whole, rule out the possibility of the match being lost.

In concluding this short section, we would like to say that for any dual positional game there is an equivalent matrix game. Therefore, finding an optimal behavior pattern in the game in the extensive form reduces to finding optimal strategies in equivalent matrix games.

Let us clarify our concept and consider the example of a somewhat simplified but interesting situation. Assume, as before, that each of the two contending lifters is entitled to one attempt at lifting or to skipping one weight. Assume, also, that the competition is started by the first lifter (I), that he is lighter than the second lifter (II), and that they both have already lifted an equal weight. In that case the tree diagram is as shown in Figure 26. Note that $+1$ denotes the first lifter's success and that -1 denotes the second lifter's success. Recall, once again, that it is natural to suppose that $p_2 < p_1$ and $q_2 < q_1$.

Let us consider the situation from the first lifter's standpoint. If lifter I decides to lift the first weight and takes it, then lifter II, who is heavier, will have to attempt the next weight. If, however, lifter I fails to take the first weight, then lifter II will obviously have to request the same weight, since the probability q_1 that he lifts it is greater than the probability q_2 that he lifts the next weight.

Let us estimate the payoff of lifter I if he decides to lift the initial weight. He will lift it with probability p_1; in that case, lifter II, being aware of this result, will decide to lift the next and greater weight, with probability q_2. Hence, the first lifter's payoff will be

$$-1 \cdot q_2 + 1 \cdot (1 - q_2) = 1 - 2q_2.$$

Lifter I will fail to lift the initial weight with probability $1 - p_1$, while lifter II will take it with probability q_1 or fail to take it with probability $1 - q_1$.

In that case, the first lifter's payoff will amount to

$$-1 \cdot q_1 + 1 \cdot (1 - q_1) = 1 - 2q_1.$$

Thus, if lifter I is going to attempt to lift the initial weight, his expected payoff $v_1(+)$ is computed by the formula:

$$v_1(+) = p_1(1 - 2q_2) + (1 - p_1)(1 - 2q_1) = 1 - 2q_1 + 2p_1q_1 - 2p_1q_2.$$

It is clear that the second lifter's payoff v_2 in that case is equal to $-v_1(+)$. Now let us assume that lifter I decides not to make the first attempt. In that case he exchanges parts, as it were, with lifter II and waits to see what the latter is going to do. If lifter II decides to try to lift the initial weight, he will take it with probability q_1, or he will fail to take it with probability $1 - q_1$. In the former case, lifter I wins with probability p_2 and in the latter, he wins with probability 1 (lifter I is the lighter of the two). Thus, if lifter II makes his first attempt, the first lifter's average payoff will amount to

$$v_1(-) = -1 \cdot q_1(1 - p_2) + 1 \cdot (1 - q_1) = 1 + p_2q_1 - 2q_1 = -v_2(+).$$

If, on the other hand, lifter II makes his second attempt (having skipped the first), he will win with probability $q_2(1-p_2)$ or lost with probability $(1-q_2)$. Thus, in this situation

$$v_1(-) = -v_1(-) = -1 \cdot q_2(1 - p_2) + (1 - q_2) = 1p_2q_2 - 2q_2.$$

It is natural to suppose that lifter II will choose the behavior that will help minimize his payoff. In other words, he will first find the minimum of the two variables, $v_2(+)$ and $v_2(-)$, and if $v_2(+) < v_2(-)$, he will make his first attempt, but if $v_2(+) > v_2(-)$, he will postpone it till the weight is increased. Let us compare these two quantities:

$$-v_2(+) = 1 + p_2q_1 - 2q_1 \quad \text{and} \quad -v_2(-) = 1 + p_2q_2 - 2q_2.$$

Let us find their difference

$$-v_2(+) - (-v_2(-)) = p_2(q_1 - q_2) - 2(q_1 - q_2) = (q_1 - q_2)(p_2 - 2).$$

Since $q_1 > q_2$ and $p_2 < 1$, we conclude that $-v_2(+) < -v_2(-)$ or, finally, $v_2(-) < v_2(+)$. Thus, lifter II should skip his first attempt, whenever lifter I does.

In this problem, the first lifter's optimal strategy is to tackle the first weight if $v_1(+) > v_1(-)$ and to pass it up and attempt to lift the next weight if $v_1(+) < v_1(-)$.

The second lifter's optimal behavior is as follows. If the first lifter skips a weight, so should the second lifter, being the heavier of the two. If, however, the first lifter attempts to take the lighter weight and fails, the second lifter should try to take the same weight; and he should skip it and attempt the next weight, if the first lifter succeeds.

Analysis of the solution to the foregoing example has introduced nothing new in the second lifter's tactics. Still, it seems to us that this analysis makes it possible to shape a lifter's tactics at important competitions, whenever he is the first to "move" or (given more than two contestants) whenever he precedes his chief rival. It is easy to see that in any situation the first lifter should be able to decide correctly whether to "move" or "pass"; at the same time, the second lifter's decision is more obvious. That was what prompted us to examine the comparatively simple tactical problem of what a contestant should do if he has to be the first to approach the weight (to "move").

7.7. Matrix games and linear programming. Let us examine a game of total conflict with an $m \times n$ payoff matrix $A = (a_{ij})$:

$$A = \begin{pmatrix} a_{11} & \cdots & a_{1n} \\ a_{21} & \cdots & a_{2n} \\ \cdots & \cdots & \cdots \\ a_{m1} & \cdots & a_{mn} \end{pmatrix}.$$

Let a probability vector $x = (x_1, x_2, \dots, x_m)$ (where $x_i \geq 0$ for all $i = 1, \dots, m$ and $\sum_{i=1}^{m} x_i = 1$) be the first player's optimal strategy. Then, whatever strategy he may resort to, the second player cannot prevent the first from winning v or even more than v, v being the yet-unknown value of the game (the reader will recall the minimax theorem).

Thus, we can write down the inequalities

$$\begin{gathered} a_{11}x_1 + a_{21}x_2 + \cdots + a_{m1}x_m \geq v, \\ a_{12}x_1 + a_{22}x_2 + \cdots + a_{m2}x_m \geq v, \\ \cdots\cdots\cdots\cdots\cdots\cdots\cdots\cdots \\ a_{1n}x_1 + a_{2n}x_2 + \cdots + a_{mn}x_m \geq v. \end{gathered}$$

We can, besides, take into account the normalizing conditions for the vector x:

$$\sum_{i=1}^{m} x_i = 1.$$

Suppose that the value of the game v is greater than zero. Otherwise, let us add to all the elements of the matrix A one sufficiently large number, thus increasing the value of the game by this number. It will not affect the optimal strategies. Then, dividing the inequalities and the normalizing condition by v, we obtain

$$\begin{gathered} a_{11}\frac{x_1}{v} + a_{21}\frac{x_2}{v} + \cdots + a_{m1}\frac{x_m}{v} \geq 1, \\ \cdots\cdots\cdots\cdots\cdots\cdots\cdots\cdots \\ a_{1n}\frac{x_1}{v} + a_{2n}\frac{x_2}{v} + \cdots + a_{mn}\frac{x_m}{v} \geq 1, \\ \sum_{i=1}^{m} \frac{x_i}{v} = \frac{1}{v}. \end{gathered}$$

Now let us introduce new variables

$$\xi_1 = \frac{x_1}{v}, \dots, \xi_m = \frac{x_m}{v}$$

and note that the first player's goal is to increase v and so decrease $1/v$. Thus we have arrived at a linear programming problem with constraints in the form of inequalities, which is to minimize the sum

$$\sum_{j=1}^{n} \xi_j = \frac{1}{v}$$

subject to the constraints

$$\begin{aligned}
&a_{11}\xi_1 + a_{21}\xi_2 + a_{31}\xi_3 + \cdots + a_{m1}\xi_m \geq 1, \\
&a_{12}\xi_1 + a_{22}\xi_2 + a_{32}\xi_3 + \cdots + a_{m2}\xi_m \geq 1, \\
&\cdots\cdots\cdots\cdots\cdots\cdots\cdots\cdots\cdots\cdots \\
&a_{1n}\xi_1 + a_{2n}\xi_2 + a_{3n}\xi_3 + \cdots + a_{mn}\xi_m \geq 1, \\
&\qquad \xi_i \geq 0 \qquad (i = 1, \ldots, m).
\end{aligned}$$

The reader has already learned how to solve such problems (linear programming problems) and will be able to find the values of ξ_j (and so of x_j too) which constitute the optimal strategy of the first player.

Note that finite games with a great number of strategies are often solved by the approximate (iterative) method. Many simple iterative procedures have now been developed, which can easily be programmed for a computer and yield optimal strategies.

And now—hockey!

7.8. Zany Zebras at Mayapple Leafs (on a hockey theme). Here is a report from the *Daily Echo* of April 31, 1989.

"As usual, at the end of the season two national hockey teams vied for the Silver Cup of the continent. It was a hard-fought match: in the first period the Zebras led 3:2, raising the score to 4:3 in the second period. The third period opened with an attack on the Zebras' goal, the Leafs equalizing the score in the middle of the period. The advantage passed to the Leafs, yet they still failed to score the deciding goal.

"Five minutes before the end the spectators began wondering at the bizarre alternation of the Leafs' fives, orchestrated by their coaches. They stopped moving in the fourth five. The first five appeared on the ice less often. But the second and third fives kept alternating at a mad pace. The astonished spectators, already beginning to boo, suddenly broke into loud cheering: ninety seconds before the end, the deciding goal was scored! The Mayapple Leafs triumphed with a score of 5:4, winning the challenge cup for the fourth time in a row.

"At the press conference after the match, the Leafs' coach said that they owed their victory entirely to a young mathematician who had started working with the team toward the end of the season. His words caused a sensation. Cameras clicked, flash bulbs blazed, several newsmen put their microphones in front of the cause of the sensation, asking him to explain how mathematics could have helped win such a momentous contest.

"Here is what the consulting mathematician told them (we print a somewhat simplified account of what took place on the bench in the final ten minutes of the match):

"'We were in the lead but somehow failed to realize our advantage. Our fives—all four of them—clearly outplayed the opponents, but somehow nobody scored. I decided to keep reshuffling the fives and move them in at random, without waiting to see what the opposition would do. As a matter of fact, the Zebras were much more tired than our players, and the slow switching of fives—each coach letting out now one five, now another—enabled the Zebras to slow down our attacks. I drew up a rectangular table, assigning the rows to our fives and the columns to theirs. We had four fives while the Zebras had three fives left, after two players from the fourth five had sustained injuries. Let us denote our fives by α_i and their fives by β_j. At the intersection of the row corresponding to our five α_i and the column of their five β_j, I entered the estimate a_{ij} (on a ten-point scale) of the advantage of α_i over β_j.

"'I got the estimate a_{ij} by averaging the individual estimates I had picked up from our coaches who formed them based on previous matches.

"'Recently I've taken to carrying about my microcomputer, so I was able to put together on the spot the following table (I should like to stress right away that I could have made up an analogous table in which I should have entered, as a_{ij}, probability p_{ij} that the five α_i would win over the five β_j):

"Zany Zebras"

$$\text{"Mayapple Leafs"}\quad \begin{matrix} \\ \alpha_1 \\ \alpha_2 \\ \alpha_3 \\ \alpha_4 \end{matrix} \begin{matrix} \beta_1 & \beta_2 & \beta_3 \\ \end{matrix} \begin{pmatrix} 8 & 2 & 4 \\ 6 & 3 & 6 \\ 2 & 7 & 6 \\ 1 & 7 & 3 \end{pmatrix}.$$

"'After that I set an equivalent linear programming problem with constraints (see 7.7)

$$8x_1 + 6x_2 + 2x_3 + x_4 \geq 1,$$
$$2x_1 + 3x_2 + 7x_3 + 7x_4 \geq 1,$$
$$4x_1 + 6x_2 + 6x_3 + 3x_4 \geq 1$$

and the linear function

$$F(X) = x_1 + x_2 + x_3 + x_4 = \frac{1}{v}$$

to be minimized. You will recall that here v is the yet-unknown value of the matrix game I have constructed.

"'I won't make you listen to a detailed account of the solution of the linear programming problem that I obtained—you can easily solve it by the simplex method—and will just give you the answer. The optimal solution is:

$$x_1^0 = \frac{1}{32}, \quad x_2^0 = \frac{3}{32}, \quad x_3^0 = \frac{3}{32}, \quad x_4^0 = 0, \quad \text{and } \frac{1}{v} = \frac{7}{32}.$$

"'Thus, the value of the game turned out to be $\frac{32}{7}$, and the optimal probabilities of the pure strategies being used in our coaches' mixed strategy (i.e., the optimal strategy of my team) was defined as follows:

$$p_1^0 = \frac{1}{32} \cdot \frac{32}{7} = \frac{1}{7}, \qquad p_2^0 = \frac{3}{32} \cdot \frac{32}{7} = \frac{3}{7},$$
$$p_3^0 = \frac{3}{32} \cdot \frac{32}{7} = \frac{3}{7}, \qquad p_4^0 = 0 \cdot \frac{32}{7} = 0.$$

That was why we let out the first five less often and kept the fourth five altogether off the ice. Since the opposition did not use its optimal strategy of finishing the game with two fives and their third five on the bench, we, naturally enough, turned our advantage into the deciding goal.' "

A speech like this ought to set coaches thinking over the problems that arise at hockey matches.

7.9. How to form a swimming team. One of the first suggestions on how to use the solution algorithm of the assignment problem (see 6.1) in order to form a team is set forth in a paper by R. E. Machol, entitled *An application of the assignment problem* (Oper. Res. **18** (1970), 569–760). It discusses the formation of a medley relay team including several members able to swim more than one stroke.

In a medley relay, each member of a four-strong team swims a different stroke for each leg in a definite succession. The backstroke is swum first, then the breast stroke, butterfly, and freestyle.

A more involved problem, concerned with selecting a basic team from among 38 candidates, was considered in 1972. The paper mentioned above sets out a logical, if rather complicated, procedure for determining the coefficients of the objective function (in our notation, the elements of the point matrix Γ). First the coach draws up a list of the indicators (physiological, intellectual, psychological, etc.) that a candidate should have to be able to occupy at training sessions a certain place in the relay. From this list he selects n indicators excluding one another as far as possible, yet describing the candidate well enough to define the "independent" indicators of his ability.

Then each coach assesses the significance of each indicator, attaching to it a certain "weight" (a numerical factor) taking account of the place occupied by the player during practice. Next, by some method of appraisal, used in examination by experts (see 4.1), a collective weight factor is calculated. The final result is an $n \times m$ "weight" matrix $\Lambda = (\lambda_{ij})$. Its element λ_{ij} is the weight assigned by the coaches to the ith indicator $(i = 1, \dots, n)$, provided that the swimmer is in the jth place $(j = 1, \dots, m)$.

At the next stage, the decision maker (e.g., the head coach) defines the significance of each of the m places in the team and, resorting again to the weight factor technique, assigns to the jth place a weight μ $(j = 1, \dots, m)$. After this, the coaches evaluate each of the k swimmers. To this end, they construct a $k \times n$ weight matrix $A = (a_{li})$, a_{li} being the weight of the lth swimmer with respect to the ith indicator. Multiplying $A_{k\times n}$ by $\Lambda_{n\times m}$ gives a matrix $\Gamma(\gamma_{lj})_{k\times m} = A_{k\times n}\Lambda_{n\times m}$. Its element $\gamma_{ij} = a_{l1}\lambda_{1j} + a_{l2}\lambda_{2j} + \cdots + a_{ln}\lambda_{nj}$ estimates the worth to the team of the lth swimmer occupying the

jth place $(l = 1, \ldots, k;\ j = 1, \ldots, m)$.

The estimate of the contribution made by the lth swimmer being in the jth place is given by the indicator $\delta_{lj} = \gamma_{lj}\mu_j$. We arrive at a matrix $\Delta_{k\times m} = (\delta_{lj})$. In conclusion, the already-familiar assignment problem is solved by the linear programming method. The variables $x_{lj} = 0$ or 1 are introduced (depending on whether or not the lth swimmer is assigned the jth place); the system of constraints is written as

$$\sum_{l=1}^{k} x_{ij} = 1 \qquad (j = 1, \ldots, m),$$

$$\sum_{j=1}^{m} x_{lj} = 1 \qquad (l = 1, \ldots, k),$$

and, provided they are satisfied, the objective function—the effectiveness of the team's performance—is maximized

$$F(X) = \sum_{l=1}^{k}\sum_{j=1}^{m} \delta_{lj}x_{ij}.$$

In specific cases, this problem can be solved by the simplex method (see 6.8) by means of a computer.

8

Organizing Competitions is an Operations Planning Problem

Competitions in various sports are organized following different principles or planning systems, the better-known of which are the *Olympic* (*cup*) and *round robin*. Under the Olympic system, more contestants can initially be involved in competitions compared with any other system.

In a round robin, each contestant is matched against every other contestant, which makes the results less dependent on chance. When conducted in two rounds, the results of such a tournament are still more accurate.

8.1. The Olympic system. In tournaments carried out under the Olympic system, a contestant (or team) is out of the running after the first defeat. Afterwards, such contestants may claim an $(n+1)$th place (given $2n$ participants), if an additional group is set up for them (usually, to contest places from the $(n+1)$th to the $2n$th).

Contestants draw their numbers. The contestants $A, B, C, \ldots$ are entered in the draw sheet according to the numbers drawn. If the number of contestants is $n = 2^k$—i.e., equal to a power of two—then all contestants take part in the first round. If, however, n is not a power of two, then some of the contestants enter the competition in the second round.

The number m of those entering the competition after the first round is the difference between 2^k (the power of two, closest to the number n of contestants and larger than n), and the number n of contestants: $m = 2^k - n$, the number of the pairs playing in the first round being $n - 2^{k-1}$. If the number of contestants is $n = 2^k$, then all matches will be played in k rounds. But if $2^{k-1} < n < 2^k$, then the winner has to play $k+1$ (or k) matches.

The numbers entering in the second round (when $2^{k-1} < n < 2^k$) are selected in various ways. For example, they may be numbers from the top and bottom of the draw sheet. Another way is for the draw sheet to contain 2^k positions of which only n positions are filled by contestants. The vacant positions are filled later on by additional contestants or are left vacant and placed beside the positions filled by major contestants. To prevent the latter from knocking one another out in the early rounds, they are "seeded" all over the draw sheet.

For example, in tennis tournaments the better players $(2, 4, 8, 16, \ldots)$ do not take part in the draw. They are assigned positions in the different halves, quarters, eighths, and sixteenths of the draw sheet. Then the two best contestants draw for the first and last positions in the draw sheet; the next two draw for the positions in the middle (32 and 33, given 64 contestants), and the next four draw for numbers in the middle between those already drawn (16, 17, 48, 49, given 64 contestants).

There are also other ways of organizing tournaments under the Olympic system **[14]**.

The accuracy of the results of competitions conducted under the Olympic system stems from a principle known in mathematics as *transitivity*, namely, it is considered a priori that if A plays (say tennis or chess) better than B and B plays better than C, then A plays better than C.

However, the transitivity principle often does not hold, especially in tennis. In the 1984 French Open, for example, Ivan Lendl, then third in the world rankings, beat the world's number one, John McEnroe. At Wimbledon McEnroe beat another American, Jimmy Connors, second in the world rankings. And Connors (also at Wimbledon) beat Lendl. This also happens among chess players. For example, Robert Fischer scored over Bengt Larsen on the basis of individual matches, Larsen scored over Yefim Geller on the basis of games, and Geller scored over Fischer on the basis of points.

Violations of the transitivity principle were already discussed in connection with judging in figure skating (see 4.4).

Still, the Olympic system is widely applied at major international tennis competitions, such as the Davis Cup and Grand Prix tournaments, Australian, U.S., and French opens, and the Wimbledon championships.

The Olympic system is rarely used at chess tournaments where preference is given to *round robin* and the *Swiss System.*

8.2. Round robin. This system is widely practiced in soccer, hockey, basketball, and tennis. Tournaments conducted in two or three rounds, have highly accurate results. The results of a round robin tournament can be represented both as a summary table and as a graph. The vertices of a graph (n in number) correspond to contestants. Each pair of vertices is joined by an edge, so that there are altogether $n(n-1)/2$ edges. Such a graph is called *complete.* If contestant A_i wins a match against A_j, the edge joining the respective vertices is supplied with an arrow directed from A_i toward A_j). In case of a draw, the edge remains nonoriented (left without an arrow).

There are no draws in tennis (as, incidentally, in basketball and volleyball). A round robin in these games is a "no-compromise" tournament. A complete oriented graph corresponds to such a tournament. For example, all the possible graphs of no-compromise tournaments involving four contestants are represented in Figure 27. Such tournament graphs are studied in

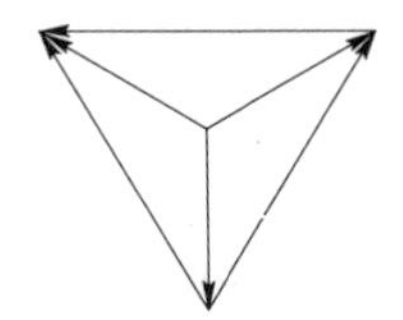
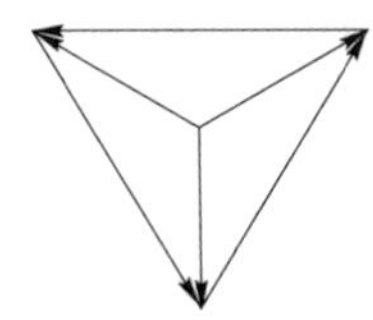
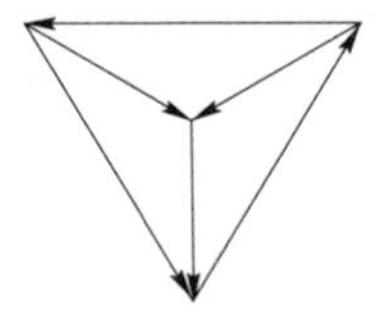
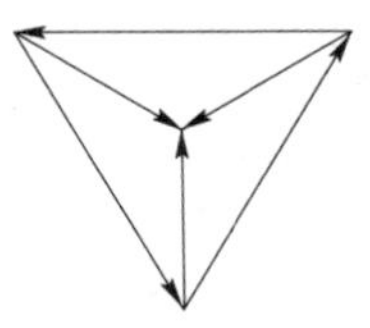

FIGURE 27

graph theory, a mathematical discipline closely associated with combinatorics (see, e.g., **[21]**).

In chess round robins the schedules are strictly adhered to. Players in each round are assigned numbers and choose colors by lot. All calculations are made following the rules proposed by the German chess player J. Berger.

Here is, basically, what these rules are about. We assume that there is an even number n of contestants. Let us consider a pair of contestants. If their drawn numbers s and t differ from n, then the number of the round in which they are matched is $N_{st} = s + t - 1$ given $s + t < n$, or $N_{st} = s + t - n$ given $s + t > n$. If the sum $s + t$ is an odd number, the player with the smaller number plays with white, and if the sum is even, then his opponent plays with white. For example, in a tournament with eight contestants ($n = 8$) number two ($s = 2$) plays number five ($t = 5$) with white in the sixth round ($2 + 5 = 7$ is an odd number smaller than eight; $2 + 5 - 1 = 6$); and number three plays number seven with black in the second round ($3 + 7$ is an even number greater than eight; $3 + 7 - 8 = 2$). The round in which player number n is matched with player number l is $2l - 1$ (f $2l < n$) and $2l - n$ (if $2l > n$). He plays numbers 1 through $n/2$ with black, and the other numbers with white.

If n is an odd number, then a dummy player is introduced and the problem of drawing up the schedule is reduced to the even-numbered case $n + 1$. Games with a dummy partner are not played, and the real partner is free in that round.

Note that when n is even, contestants with numbers from 1 to $n/2$ play an extra game each with white (because they play number n with white). For this reason the more desirable numbers are those not exceeding $n/2$.

Tennis round robins are also strictly regulated today.

Round robins are a source of many interesting mathematical problems.

Here is the simplest of them. In a tennis round robin two players have an equal number of victories. Can all participating players be ranked according to their playing strength? It turns out that they cannot, since there will always be among them three players, A, B, and C, such that A beat B, B beat C, and C beat A (there is an oriented cycle in the relevant graph). Indeed, assume that A and B have an equal number k of wins and that A beat B. Suppose that B boasts victories over the players $C_1, \dots, C_k$. Then there must be at least one among them who has beaten A. Otherwise, A would have had not less than $k + 1$ victories.

Some intriguing situations associated with chess battles are described by E. Ya. Ghick in **[24]**. We shall cite two of them.

1. A chess tournament has n contestants. What may be the maximum difference in points scored by two players succeeding each other in the final score table?

Let us assume that the largest gap is between the players in positions s and $s+1$. The players in the first s positions have played $s(s-1)/2$ games with one another and scored altogether as many points. Besides, they have played $s(n-s)$ games with the players in positions from $s+1$ to n, scoring not more than $s(n-s)$ points. Thus, the total number of points scored by the players in the first s positions does not exceed $\frac{s(s-1)}{2}+s(n-s)=\frac{(2n-s-1)s}{2}$. Since the player in position s is the last of the first s players, he has scored not more than $\frac{1}{s}\frac{(2n-s-1)s}{2}=\frac{2n-s-1}{2}$ points.

The players in positions $s+1$ to n have played $(n-s)(n-s-1)/2$ games with one another and scored as many points. Since the player in position $s+1$ is the first of them, he has scored not less than $\frac{1}{n-s}\frac{(n-s)(n-s-1)}{2}=\frac{n-s-1}{2}$ points.

Hence, the maximum gap between the players in positions s and $s+1$ does not exceed $\frac{2n-s-1}{2}-\frac{n-s-1}{2}=\frac{n}{2}$. Such a gap may occur, for example, when the champion has beaten all his opponents and scored $n-1$ points, while the games played by the rest of the players with one another were all drawn, so that they all scored $n/2-1$ points each. Furthermore, $(n-1)-(n/2-1)=n/2$.

At major chess tournaments such a gap is unlikely. Nevertheless, in a unique episode in the history of class Alechine, at a tournament in 1931 at Bled, got 5.5 points ahead of his rivals.

2. Sometimes it may be necessary to renumber the players after a tournament to avoid the winners being placed immediately after the respective losers.

Let us see if such a reordering is possible. We shall proceed by mathematical induction. Ordering for two players is obvious. Assume that it is done for any tournament with m players. Let us turn to the result of a tournament with $m+1$ players. Let us choose any m players and order them as required (we can do it by inductive assumption). Then let us find out how the $(m+1)$st player fared against the first. If he won or it was a draw, we shall put him first. If he lost, we shall find out how he managed with the second player, and so on. If, during this search, we find the player whom the $(m+1)$st beat or the game was drawn, then we shall put the $(m+1)$st before that player. If we find that the $(m+1)$st has lost to everybody, then we shall put him last.

8.3. The Scheveningen[23] system and Latin squares. The Scheveningen system for competitions between teams gives rise to numerous mathematical problems. In a tournament between two teams, every member of one team meets every member of the other. The question is how to draw up the schedule. If each team has n members, the match is carried out in n rounds.

[23] After the Dutch town of Scheveningen.

	1	2	3	4
1	(1)	2	(3)	4
2	3	(4)	1	(2)
3	(2)	3	(4)	1
4	4	(1)	2	(3)

FIGURE 28

	1	2	3	4	5	6
1	1	(2)	3	(4)	5	(6)
2	(2)	3	(4)	5	(6)	1
3	3	(4)	5	(6)	1	(2)
4	(4)	5	(6)	1	(2)	3
5	5	(6)	1	(2)	3	(4)
6	(6)	1	(2)	3	(4)	5

FIGURE 29

A schedule for $n = 4$ is shown in Figure 28. The members of team I are assigned to the rows and the members of team II are assigned to the columns of the square. The intersection of the ith row and jth column indicates the round in which the corresponding players take part. The color of the pieces with which a member of team II plays is determined by the color (shading) of the cell.

A schedule for $n = 6$ is shown in Figure 29. In each row and column of the square setting the schedule of rounds for $n = 4$, each figure from 1 to 4 appears exactly once. In each row (column) of a square for $n = 6$, each figure from 1 to 6 appears exactly once. It is similar for any n.

Squares with such an arrangement of numbers are called *Latin squares.* Until recently, Latin squares were of interest only to puzzle lovers and some mathematicians. Latin squares became known mainly thanks to the problem, proposed in 1782 by Leonhard Euler (1707–1783): Among 36 officers, there are six each of six different ranks, from six regiments. Can they be ranged in a square formation so that officers of each rank and regiment could be found in every line and column?

It was proved only in 1901 that such an arrangement was impossible. Nevertheless Latin squares (in working with them, Euler used letters of the Latin alphabet, rather than numbers) found numerous applications, specifically in combinatorial problems. In the late 1960s, they came to be applied in message encoding theory (see **[23]**).

We have also seen that Latin squares have a direct bearing on the organization of competitions. So let us talk about them some more.

So, a *Latin square* is a matrix of order n, whose elements are the numbers $1, 2, \ldots, n$, each of them occurring exactly once in each row and each column.

Two matrices of order n are said to be orthogonal if, when superimposed on each other, there occurs a set of all ordered pairs (i, j), $i = 1, \ldots, n$; $j = 1, \ldots, n$.

Our example of orthogonal Latin squares (matrices) of order $n = 3$ is provided by

$$\begin{pmatrix} 1 & 2 & 3 \\ 2 & 3 & 1 \\ 3 & 1 & 2 \end{pmatrix}, \qquad \begin{pmatrix} 1 & 2 & 3 \\ 3 & 1 & 2 \\ 2 & 3 & 1 \end{pmatrix}$$

When they are superimposed, a new square appears,

$$\begin{pmatrix} (1, 1) & (2, 2) & (3, 3) \\ (2, 3) & (3, 1) & (1, 2) \\ (3, 2) & (1, 3) & (2, 1) \end{pmatrix},$$

called a Greco-Latin or Eulerian square. Its elements are ordered pairs of numbers (i, j).

Let the first figure in each pair of numbers be a rank (e.g., 1–lieutenant, 2–captain, 2–major, etc.), and the second figure be a regiment number. We obtain a Eulerian square with the required arrangement of nine officers of different ranks, from three regiments.

The following matrices are also orthogonal:

$$A = \begin{pmatrix} 1 & 1 & 1 & \cdots & 1 \\ 2 & 2 & 2 & \cdots & 2 \\ 3 & 3 & 3 & \cdots & 3 \\ \cdots & \cdots & \cdots & \cdots & \cdots \\ n & n & n & \cdots & n \end{pmatrix}, \qquad B = \begin{pmatrix} 1 & 2 & 3 & \cdots & n \\ 1 & 2 & 3 & \cdots & n \\ 1 & 2 & 3 & \cdots & n \\ \cdots & \cdots & \cdots & \cdots & \cdots \\ 1 & 2 & 3 & \cdots & n \end{pmatrix}.$$

It is easy to see that a matrix of order n is a Latin square if and only if it is orthogonal with respect to both A and B. Neither of the latter, however, is a Latin square.

The problem of whether there can exist orthogonal Latin squares of any order remained unsolved for over two hundreds years. In 1779, Euler himself formulated a conjecture that there cannot exist orthogonal Latin squares of order $n = 4k + 2$ $(k = 0, 1, 2, \ldots)$. It was only in 1901 that Terry proved the truth of Euler's conjecture for $n = 6$ (i.e., $k = 1$; it is obvious for $k = 0$) by exhaustion. Then in 1959 Parker found a Eulerian square of order $n = 10$ (i.e., for $k = 2$), and finally in 1960 he showed, jointly with Bose and Shrikhande, the existence of orthogonal Latin squares of any order n, except for $n = 2$ and $n = 6$.

Let us now go back to the Scheveningen tournament schedule for $n = 6$. It can be seen that, even though each player plays an equal number of games with white and black, each team plays all games in a round with the same color, which is undesirable (the team playing with the white pieces is in an advantageous position).

Therefore, the problem is to draw up a schedule such that:

	1	2	3	4	5	6
1	1	2	3	4	5	6
2	2	3	6	5	1	4
3	3	6	2	1	4	5
4	4	1	5	2	6	3
5	5	4	1	6	3	2
6	6	5	4	3	2	1

FIGURE 30

(a) all participants play an equal number of games with white and black;

(b) in each round, each team plays an equal number of games with white and black.

Naturally, there has to be an even number n of players. Let us examine a pair of orthogonal Latin squares, A and B, of order n. We superimpose B upon A and shade all cells of A on which fall even numbers from the square B. The rest of the cells in A are left white.

Since half of the numbers in each row and column of A are even, we can see at once that the condition (a) is satisfied. A and B being orthogonal, to each n identical numbers in A correspond half of the even and half of the odd numbers in B. Thus the condition (b) is also satisfied.

Note that while no orthogonal Latin squares exist for $n = 6$, this does not mean yet that the schedule we are seeking cannot be made. Indeed, a series of such schedules were found by exhaustive search carried out by means of a computer. One of them is shown in Figure 30. Its merit is that no player plays with the same color more than two games in a row.

To give the reader an idea of the scope of the search involved in looking for the appropriate Latin squares, we shall cite Riordan **[22]**. Let us denote by l_n the number of Latin squares of order n, in which the elements of the first row and first column are arranged in the natural order: $1, 2, \dots, n$. Then then number L_n of the different Latin squares of order n will be

$$L_n = n!(n-1)!l_n,$$

since arbitrary permutation of rows (columns) does not affect the characteristics of a Latin square.

Furthermore, the values of l_n for $n \leq 7$ are as follows:

n	2	3	4	5	6	7
l_n	1	1	1	56	9408	16942080

9

Classifications in Sports

9.1. Classification principles. Athletes are classified according to their strength, based on estimates which may be expressed by the number of points scored at tournaments or by a *rating*, which is a conventional numerical coefficient.

In gymnastics, figure skating, weight lifting and some other sports, these estimates are expressed by the points scored at separate tournaments.

But in competitive games, such as chess, tennis, or badminton, it is more practical to use an integral estimator based on the results of a series of matches over a certain period. Such integral estimators are used in chess and tennis classifications, to name but two. By examining current classification systems, we can draw some general conclusions about the principles they are built on.

Let us take a look at the classification system based on ratings.

At this point, we shall not go into how initial ratings are awarded. We shall just assume that this question has been solved.

Consider a match between two players. Let us denote class estimators, or ratings, of players U and V by $r(U)$ and $r(V)$, respectively.

Let us introduce a variable t describing the distinction in class between U and V. It may be assumed to depend, for example, on the ratio $r(U)/r(V)$ or the difference $r(U) - r(V)$.

We shall proceed with the assumption that there is an exponential relationship between the ratio m/n of the average number of wins m to the average number of losses n, incurred by U in a series of N matches with V, and the difference between their ratings.

The legitimacy of this assumption is borne out by statistics on the results of chess, tennis, and similar tournaments, and, no less convincingly, by our own results presented below. Indeed, assume that the ratio

$$\frac{\langle U\text{'s victory}\rangle}{\langle V\text{'s victory}\rangle} = f(t)$$

(where $\langle U\rangle$, $\langle V\rangle$ are the number appearances of U, V); it is natural that the function $f(t)$ is i) increasing ii) greater than 0, and iii) $f(0) = \frac{1}{2}$. The simplest function satisfying these conditions is the exponential function.

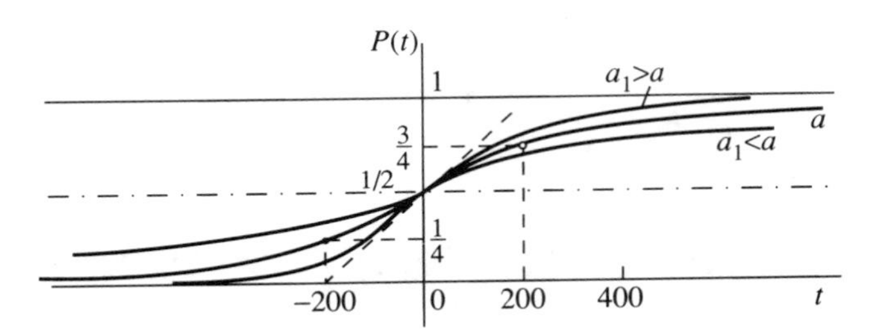

FIGURE 31

So, we assume that

$$m/n = a^t,$$

where the basis a of the exponent a^t is a number greater than one, and $t = r(U) - r(V)$. In this notation, the probability $P(U$'s victory) that U wins the match with V is equal to the ratio of the average number of matches won by U to the overall number of matches with V, i.e.,

$$P(U\text{'s victor}) = \frac{m}{m+n} = \frac{m/n}{1+m/n} = \frac{a^t}{1+a^t} = p(t).$$

It is clear that $P(U$'s victory) is a function of t. Since t is equal to the difference $r(U) - r(V)$, it is obvious that given $t = 0$ (i.e., both players are in the same class), the probability $P(0) = \frac{1}{2}$: both opponents are of equal strength.

Given an unlimited growth of t (i.e., an unlimited growth of U's rating, compared with V's), probability $P(t)$ approaches one: U is practically certain to beat V. If, however, t becomes negative $(r(U) < r(V))$ and decreases indefinitely, the probability $P(t)$ of U winning over V approaches zero.

Graphs of the functions of $P(t)$, depending on the value of the parameter a, are shown in Figure 31. The graph of the function $P_1(t) = a_1^t/(1 + a_1^t)$ is located above the graph of $P(t) = a^t/(1 + a^t)$ when $t > 0$, and below when $t < 0$, if $a_1 > a$.

All that remains to be done now is choose a numerical value for a and select a scale for estimating the difference in class between players (their rating).

In general, it is possible to explain the idea of the already classic (though not yet clearly substantiated in the literature) classification of chess players, proposed by an American professor, A. Elo, and adopted by the International Chess Federation (FIDE) in 1970.

The statistics on chess tournaments indicate that a player who is one grade higher in the chess hierarchy than his opponent wins on the average 75 out of 100 points, i.e., with probability 0.75.

In constructing the curve $p(t)$—i.e., in choosing the value of the parameter a—this fact should be allowed for as follows. Suppose that the difference between two players at successive grades of the chess hierarchy amounts to λ rating units. In other words, with $t = r(U) - r(V) = \lambda$, the probability that U beats V is equal to $P(\lambda) = 0.75$. This assumption leads to the following relation for finding a:

$$\frac{a^\lambda}{1 + a^\lambda} = 0.75,$$

or $a^\lambda = 3$. Assuming, for example, that $\lambda = 200$, we shall find from $a^{200} = 3$ that $a = 1.0055$.

Now we can compute that, given a difference in ratings $t = 5$,

$$P(5) = \frac{a^5}{1 + a^5} = 0.510;$$

for $t = 10$, $P(10) = 0.514$; for $t = 15$, $P(15) = 0.520$; for $t = 20$, $P(20) = 0.527$, and so on.

Having computed the probabilities $P(t)$ for rating differences from $t = 0$ to $t = 735$ (and over), we arrive at Elo's table (Table 10) (it is cited, for example, in **[24]**). The variable ΔK—the difference between the players' Elo coefficients in our notation is equal to the value of t. The variables h_b and h_S from Elo's table (the percentages to be collected by the player with the bigger rating and, respectively, by the one with the smaller rating) are, in effect, the already found probabilities $P(U\text{'s victory})$ and $P(V\text{'s victory}) = 1 - P(U\text{'s victory})$.

TABLE 10

ΔK	h_b	h_S	ΔK	h_b	h_S	ΔK	h_b	h_S
0–3	50	50	122–129	67	33	279–290	84	16
4–10	51	49	130–137	68	32	291–302	85	15
11–17	52	48	138–145	69	31	303–315	86	14
18–25	53	47	146–153	70	30	316–328	87	13
26–32	54	46	154–162	71	29	329–344	88	12
33–39	55	45	163–170	72	28	345–357	89	11
40–46	56	44	171–179	73	27	358–374	90	10
47–53	57	43	180–188	74	26	375–391	91	9
54–61	58	42	189–197	75	25	392–411	92	8
62–68	59	41	198–206	76	24	412–432	93	7
69–76	60	40	207–215	77	23	433–456	94	6
77–83	61	39	216–225	78	22	457–484	95	5
84–91	62	38	226–235	79	21	485–517	96	4
92–98	63	37	236–245	80	20	518–559	97	3
99–106	64	36	246–256	81	19	560–619	98	2
107–113	65	35	257–267	82	18	620–735	99	1
114–121	66	34	268–278	83	17	upwards of 735	100	0

The coincidence shows that 200 is the number adopted in the Elo System as the starting rating difference.

Let us point out, however, that the rating difference λ between players at two successive grades of the chess hierarchy can be assessed to be say 250 (or even 300) rating units, rather than Elo's 200. This would result in a smaller value of $a = 1.0044$, found from the equation $a^{250} = 3$, and, naturally, in another table, albeit similar to Elo's [**31**].

It is quite possible that the new table will prove to be better than Elo's for recalculating the player's ratings after a tournament. The only way to justify the choice of $\lambda = 200$ is to study the vast statistical material on chess battles, accumulated over the years.

Anyway, it is possible to construct an entire family of similar classifications. In particular, it may be assumed that the difference between the ratings of players in each successive pair of grades should not be constant but, rather, should be changing upward (from third grade to grand master). Granted, this will make recalculation of ratings somewhat more difficult, and so hardly practical.

But let us return to the Elo System. It concludes with a rule for recalculating ratings. This rule is formalized as the linear dependence

$$r_n = r_o + \mu(N - N_e) \tag{1}$$

expressing the new rating (at the end of a tournament) r_n through the old rating r_o (before the tournament), and through the difference between the number N of the points scored and the expected number of points N_e presumably "due" a player by virtue of his qualifications.

If the actual score N coincides with the expected N_e, the player's rating—as follows from (1)—remains unchanged: $r_n = r_o$.

If $N > N_e$, i.e., if $N - N_e > 0$, his rating will increase; if $N < N_e$, it will become less than it was before the tournament.

In the Elo System, the coefficient $\mu = 10$. Hence, if $N - N_e = 1$, it follows from the formula

$$r_n = r_o + 10(N - N_e)$$

that the rating will increase by 10. In other words, one point scored over and above the expected score will fetch 10 rating units. This is also an "imposition." One can well assume say $\mu = 15$ or $\mu = 20$.

To recalculate a rating, one has to know the value of N_e. Here is how it is found (it is always rounded off up to half a point, so that ratings are always integers).

Assume that a player A, who has a rating $r_o(A) = 2280$ is matched in a tournament with players B, C, D, E whose ratings are $r_o(B) = 2280$, $r_o(C) = 2285$, $r_o(D) = 2270$, and $r_o(E) = 2260$. Then the probability that A beats B is 0.5 (identified with the average score); that A beats C, 0.49 since $(r_o(C) - r_o(A) = 5)$; that A beats D, 0.514 (since $r_o(A) - r_o(D) = 10$); and, lastly, that A beats E, 0.527 (since $r_o(A) - r_o(E) = 20$). Hence, A can be expected to score altogether

$$N_e(A) = 0.500 + 0.490 + 0.514 + 0.527 = 2.031 \text{ points}.$$

Calculation of N_e is made easier by the introduction of the *tournament rating* $r(T)$, which is the arithmetic mean of the ratings of all participants.

It is obvious that the stronger the players, the greater the tournament rating. In general, for a tournament to be rated under the Elo System, at least two thirds of its participants should be officially rated players, and the tournament rating should be at least 2250. The unrated contestants are given an initial rating $r_{in} = 2200$.

Then, rather than finding the values of t for all matches of A with B, C, D, E, as above, it is found just once, as the difference between A's rating and the tournament rating $r_o(A) - r(T)$. It is thus assumed that he plays $n - 1$ games (given n participants) with a fictitious partner who presumably has the tournament rating $r(T)$.

For example, assuming that only A, B, C, D, E, are playing in the tournament, we shall find the tournament rating

$$r(T) = \frac{1}{5(2280 + 2280 + 2285 + 2270 + 2260) = 2275}.$$

A's rating $r_o(A) = 2280$. The difference $t = 2280 - 2275 = 5$. It was previously computed that $P(5) = 0.510$. Since A is matched with the fictitious partner four times, his expected score $N_e(A) = 0.510 \times 4 = 2.040 \approx 2$.

Assume, further, that A actually scored in the tournament $N(A) = 3.5$ points. Then $r_n(A) = 2280 + 10(3.5 - 2) = 2280 + 15 = 2295$, and so A's rating has increased by 15 units.

Thus, the Elo System is based on three assumptions:

(1) the ratio of the average number of wins to the games played exponentially depends on the difference between the players' ratings;
(2) the difference in the ratings of players of two consecutive grades of the chess hierarchy is 200 rating units;
(3) one point scored (over and above the expected score) corresponds to 10 rating units.

Thus, the Elo System has a certain theoretical statistical basis. Its validity has been confirmed by the statistical accuracy of its forecasts.

Seven years before FIDE adopted his system, Elo had already computed the ratings of prominent chess players starting with P. Morphy (a rating of 2690). A rating $r \geq 2600$ was bestowed on 28 chess players (E. Lasker, J. R. Capablanca, M. Botvinnik, M. Tal, and others).

To win the rank of grand master, for example, one must, under FIDE rules, repeat a prescribed score which depends on the tournaments' category of difficulty.

Tournaments are divided into sixteen categories of difficulty, depending on their ratings.

The first category has a rating 2251 to 2275; the sixteenth category has a rating 2626 to 2650.

For example, a would-be international master (grand master) has to score in a tenth-category tournament (with a rating of 2476–2500) at least 47% (67%, respectively) of the maximum possible number of points.

He may repeat the required score at a tournament of a different category, in keeping with a scale adopted by FIDE.

The current classification of chess players in the USSR is based on the Elo System but has certain distinctions in the way ratings (coefficients) are calculated, namely,

(1) the expected score N_e is rounded off to 0.1 of a point;

(2) tournaments are graded by difficulty on a different scale and a different number of points is required to be scored to win a rank;

(3) a player with a pre-tournament rating $r_o = 2200$ is given a new rating r_n under the rule

$$r_n = 2200 + 200N/N_M,$$

where N_M is a master's required score, and N the actual score. Under their rule a successful player can notably increase his rating. For example, if he scores as much as a master's required norm, his rating will grow by 200 units.

9.2. The international tennis classification. The international Association of Tennis Professionals (ATP) uses a mathematical method of ranking men professional tennis players, introduced in 1979.

The method is realized as a computer ranking system, Atari-ATP, which is under the jurisdiction of the ATP Men's International Professional Council. The ranking is based on the results of major international tournaments, registered by the system.

For all that it is called a mathematical method, the Atari-ATP has very little to do with mathematics as such. It is an efficient machine system for counting up points, based on certain rules and tables. It is essentially a commercial system. When all is said and done, it all hinges on money—on tournament prize funds. Let us take a look at the basic features of the system.

Ranking extends to every tennis player who has won not less than a fixed sum of money (dollars) in tournaments over the preceding twelve months. This sum is fixed by ATP. The classification is revised approximately forty times during a year. Classification according to results of the singles and doubles matches is done separately.

Every player is evaluated (given a respective position in the classification) by his point average which is equal to the ratio N/τ of the number of points N scored in tournaments to the number of tournaments τ. Furthermore, τ cannot be less than 12, even if the actual number of the tournaments is less. If, however, the player took part in more than twelve tournaments, then τ is found from a special table.

Professional tennis tournaments are divided into three classes (series), namely, star, qualifying (assigned or called by the Council), and satellite tournaments.

Star tournaments, in their turn, are subdivided into categories according to their "star content". The star category of a tournament is defined by two parameters—its prize fund and the number of participants (i.e., the size of the circuit). The players are matched according to the Olympic system. The number of the stars is determined from special tables.

The computer system calculates points and awards rankings in conformity with stringent rules.

Besides the points given to a player for the place he has won in a certain category of tournament, he is additionally given bonus points for personal victories and, separately, for taking part in qualifying tournaments (by assignment).

Every participant of a qualifying tournament who successfully advances to the basic circuit (i.e., the star circuit) is given a point. He receives another point each time he beats a player ranked in the first 150. However, the total of points scored in a qualifying tournament should not be more than three.

Participants in doubles events are given points by the same method as those playing in singles events.

Participants in a satellite tournament are given points on the basis of a special table.

Bonus points in a satellite tournament do not depend on its category and are given only to the singles champion, singles finalist, and the doubles champion.

In a masters' satellite tournament, bonus points are given to the singles and doubles champions, finalists and semi-finalists, depending on the category of the tournament, based on a special table.

Such are the chief characteristics of the ATP ranking system, which can be provisionally described as a "point" system.

9.3. Domestic tennis ranking systems. Currently in the Soviet Union there are two rival systems of ranking tennis players. According to one of them (suggested by E. V. Tsarev), ranking is based on awarding every player a rating that is recalculated after each tournament. In this respect, this rating system is similar to the Elo system of classifying chess players. Even so, Tsarev's system rests on a different theoretical–statistical basis and also differs substantially from the Elo System in its methods for recalculating ratings, awarding initial ratings, and data processing.

The rating system has currently been adopted for setting up children's and junior classifications.

The system has ample software for working with a computer in a conversational mode and a capacious listing of tennis players. Post-tournament ratings can also be recalculated manually.

Under the other classification system (proposed by A. I. Naumko and G. A. Kondratieva) ranking is based on calculating the points scored by players in competitions. This "point system" (effective for tennis players older than 18) is similar, in a way, to the ATP system described above. At the same time, it is fundamentally different from it in spirit. Under the ATP system, the points given to players are, in the final analysis, determined by a tournament's prize fund. Here, points are calculated based on other parameters, namely, the rank of the competitions—their significance for the sport, the desire to upgrade the players' performances, and so on. The system introduces changes in the rankings twice a year—after the winter cycle of competitions (on covered courts) and after the spring–autumn cycle (on open courts). The points are calculated by means of a computer or manually.

10

Conclusion

10.1. Do not get carried away. Every model of a real object has relative value. It all depends on how accurately it describes the object. This largely refers to mathematical models applied in operations research and decision problems. The models of many sporting situations, discussed in this book, should not be taken literally, as guidelines for immediate action. As a rule, a model cannot reflect every feature of the situation. In the first place, it does not reflect enough—if at all—the players' psychological and physiological makeups, the coaches' experience and intuition, and, generally, what is known as "common sense". In principle, by devising a more complex model, some of these factors can be accounted for (see, for instance, modeling of tennis, 3.12). However, one needs to remember that in building a model, choosing the effectiveness criterion, and hence justifying one's choice of decision, the arbitrary approach cannot be avoided. Mathematical methods in operations research do not eliminate, but merely show where arbitrary approach was exercised. It would be hard to name a method in operations research that could absolve the decision maker from the need to review the situation as a whole and use common sense along with the results of mathematical calculations. Such calculations may yield perfectly unexpected, important information. These considerations (expressed by many noted experts in operations research) can be illustrated by an example drawn from the theory of antagonistic games. There both players are assumed to know what the opponent's strategies are. The only thing they do not know is which of the strategies are going to be selected by each side. In fact, however, a player has no advance knowledge of all the moves his opponent may make. Therefore, a player should choose a strategy that his opponent least expects—one not included in the list of strategies known to him. Nevertheless, the theory of games is of great informative and practical value since it makes it possible to choose a decision, being guided by the results of a mathematical study of the model of a game. To sum up, our prescription is: a mathematical model plus common sense.

Certain sections of this book set forth some methods, instructions, and formulas. Using them, the reader (even without delving deep into the mathematical fabric of the text) can carry out elementary "rough" calculations, using their own input data.

10.2. Inducements to research. Early mathematical works leading to quantitative estimates in the area of recreation and sports did not rely on operations research. They dealt mostly with statistical problems, determining, for example, the frequency of interceptions and lost balls in basketball, the frequency of points being won in tennis, profits on tourism, and so on.

Only at the start of the 1970s did operations research begin to penetrate more vigorously in the new areas of organization of recreation, tourism, and sport. This process was promoted, to a certain extent, by the changing circumstances of large numbers of people in industrialized countries. As a result of a shorter work week, better living conditions, improved means of communication, and so on, leisure time increased. Thus, according to some sources,[24] leisure time accounted for 27% of the aggregate active time in 1910, 34% in 1950, and is expected to reach 38% in 2000.

These issues are dealt with in a number of recent papers which are mentioned in **[11]**. The first attempt to apply operations research to tourism dates back to 1960; it was first applied to sports in 1954.[25]

We are certain that an intelligent use of scientific method by athletic coaches can materially improve both individual and team performance. The authors of **[11]** write in this connection on p. 655 that upon analyzing competitive sports in terms of operations research one can draw the following conclusions: "the data banks are in a comparatively good condition and contain the latest information (people show keen interest in these data); the actions are repeated, so that the process can be observed repeatedly under roughly identical conditions; the rules of the games are 'rigorously defined,' ruling out any possibility for 'altering the play' in full swing; the leadership in this area must be sufficiently open to technical innovation, since coaches are actively seeking ways to provide for so-called 'competitive odds.' Presumably this search for competitive odds is just why many particularly successful applications of operations research to sports have remained zealously guarded secrets to this day."

10.3. A brief survey of applications. Being profitable, baseball has long since been an object of attention for sport and business interests. A vast volume of statistics has been accumulated, enabling experts to draw conclusions about the quality of a team's performance (the average number of successful pitches, depending on the pitcher's and catcher's proficiency, the law of distribution of hits, and so on). A simulation model of baseball was constructed with the help of the probabilistic Monte Carlo method.

Soon after, mathematical methods was applied to football. One paper[26]

---

[24] M. A. Holman, *A national time budget for the year* 2000, Sociology and Social Res. **42** (1961), no. 1.

[25] C. M. Mottly, *The application of operations research method to athletic games*, Oper. Res. **2** (1954), no. 3.

[26] W. Carter and R. E. Machol, *Operation research on football*, Oper. Res. **19** (1971), 541–544.

contains the analysis of 8,373 games in 56 rounds, including the U.S. National Football League table. It supplies important recommendations on offensive strategy.

It has been proved that an optimal strategy in a football championship may also include defeat in certain matches. Such a situation occurs when a team having secured a place in the Major League is to play yet another match in its own (lower) league. If it wins the match, it will have to face a formidable opponent in its first round in the Major League; but if it loses, it will play a weaker opponent. Such situations can be described with the help of Markov chains (see 3.14). Analysis of these situations can suggest when it is better to strive for victory and when to put up with defeat. For example, authors happened to observe a similar situation at some tennis competitions where a player would rather lose (or not play) in the first round, so as to get into the "consolation" part of the tournament for the weaker players and be able to score enough points to confirm his rating.

There are also works on how to form the basic football team, define the number of spare players, optimize the age pattern, determine replacement cycles, etc.

Recommendations are available on how to draw up an optimal weekly training program for pentathletes.[27]

The model included, as the objective function, a linear dependence on the results in each part of the pentathlon. The constraints were also linear dependences, and extended to the total time (a week) of practice in all five sports; the amount of speed training was not to be less than that of endurance training; the amount of general physical conditioning was to be greater than that of technical drill, and so on. The model was analyzed by linear programming methods.

The reader must have already taken note of the great scope of problems successfully tackled by linear programming. This explains the particular attention given in this book to some theoretical aspects of linear programming.

Problems concerned with devising an optimal strategy occur in individual as well as team competitions.

We have already described a mathematical model of weight lifting (see 7.6). This purposely simplified model assumed that each contestant was allowed to lift the weight just once and to skip the next (or the initial) weight just once. Within the framework of the model of the contestants' optimal strategies were set forth. As was noted, a similar technique can be applied to analyze a real situation, where each contestant is entitled to three attempts to lift the barbells.

Using approximately the same methods, one can analyze situations arising at high-jumping and pole-vaulting competitions, where each contestant may (a) start with any height not lower than the fixed "qualifying" height, and (b) make three attempts to clear each successive height. Having taken a certain "kick-off" height (of his choice), the athlete has the crossbar raised, and so on. He is credited with the topmost of the heights he took, all preceding attempts

[27] S. P. Landany, *Operation of pentathlon training plans*, Management Sci. **21**, no. 10.

being disregarded. An athlete who starts from a greater height husbands his strength, increasing the probability of his taking the next height. But if he fails to take it, his result is counted as zero. It is possible to find the probabilities of an athlete's expected result, depending on the starting height, and suggest the optimal starting height.

10.4. The real conclusion. Well, we are coming to the end of this book, having cautioned the reader against excessive optimism and simultaneously engendering it by setting forth the successful applications of mathematical methods in sports.

Application of mathematics, specifically of operations research, to sports is a field still waiting to be paid due attention by mathematicians and sports officials alike.

Athletic competitions provide the researcher with a wealth of material that is registered by coaches, carefully preserved and continuously piled up. There is plenty of opportunity out there to experiment, to test mathematical models and optimal strategies in situations occurring in sports. Only a tiny part—quite possibly not the most intriguing one—of the problems arising in sports has been described in the pages of this book. These authors are hopeful that with the interested reader's assistance the field of "mathematics and sports" may be enriched with fresh fascinating studies and results.

References

1. E. S. Ventcel, *Operations research*, Sovetskoe Radio, Moscow, 1972. (Russian)
2. ——, *Operations research: problems, principles, methodology*, "Nauka", Moscow, 1980. (Russian)
3. E. S. Ventcel and L. A. Ovcarov, *Probability theory*, "Nauka", Moscow, 1969. (Russian)
4. N. N. Moiseev, *A mathematician is asking questions*, "Znanie", Moscow, 1974. (Russian)
5. ——, *Mathematics becomes an experiment*, "Nauka", Moscow, 1979. (Russian)
6. F. I. Karpelevich and L. E. Sadovskiĭ, *Elements of linear algebra and linear programming*, "Nauka", Moscow, 1967. (Russian)
7. B. V. Gnedenko and A. Ya. Khinchin, *Elementary introduction to probability theory*, "Nauka", Moscow, 1983. (Russian)
8. A. Kaufmann and R. Faure, *Invitation à la recherche opérationnelle*, Paris, 1963.
9. A. N. Kolmogorov, I. G. Zhurbenko, and A. V. Prokhorov, *Introduction to probability theory*, Bibl. "Kvant", 23, "Nauka", Moscow, 1982. (Russian)
10. J. G. Kemeny and J. L. Snell, *Finite Markov chains*, Undergrad. Texts Math., Van Nostrand, Princeton, NJ, 1960.
11. J. J. Moder and S. E. Elmaghraby (eds.), *Handbook of operations research*, Van Nostrand, New York, 1978.
12. *Track and field handbook*, Fizkultura i Sport, Moscow, 1983. (Russian)
13. I. V. Absaliamova and E. V. Bogdanova, *Figure skating. Comments on judging*, Fizkultura i Sport, Moscow, 1981. (Russian)
14. *Tennis competition rules*, Fizkultura i Sport, Moscow, 1980. (Russian)
15. B. G. Litvak and A. L. Sadovskiĭ, *Some game models of expert procedures*, Moskov. Inst. Inzh. Zheleznodorozh. Transporta Trudy 640 (1979), 66–68. (Russian)
16. A. L. Sadovskiĭ, *Game approach to the organization of expert examination procedures*, Siberian Conference on Reliability of Scientific-Technological Forecasts, book II, Novosibirsk, 1981. (Russian)

17. S. Karlin, *Mathematical method and theory in games, programming and economics*, vols. I, II, Addison-Wesley, Reading, MA, 1959.
18. R. D. Luce and H. Raiffa, *Games and decisions: Introduction and critical survey*, Wiley, New York, 1957.
19. N. N. Vorobev and I. N. Vrublevskaya (eds.), *Theory of games: Positional games*, "Nauka", Moscow, 1967. (Russian)[28]
20. A. D. Myshkis and L. E. Sadovskiĭ, Bibl. "Kvant", 6, "Nauka", Moscow, 1976. (Russian)
21. O. Ore, *The theory of graphs*, Amer. Math. Soc. Colloq. Publ., vol. 38, Amer. Math. Soc., Providence, RI, 1962.
22. J. Riordan, *An introduction to combinatorial analysis*, Wiley, New York, 1958.
23. M. N. Arshinov and L. E. Sadovskiĭ, *Codes and mathematics*, Bibl. "Kvant", 30, "Nauka", Moscow, 1983. (Russian)
24. E. Ya. Gik, *Chess and mathematics*, Bibl. "Kvant", 24, "Nauka", Moscow, 1983. (Russian)
25. Yu. V. Prokhorov and Yu. A. Rozanov, *Probability theory: Basic concepts, limit theorems, random processes*, "Nauka", Moscow, 1967; English transl.Grundlehren Math. Wiss., Band 157, Springer-Verlag, New York, 1969.
26. S. A. Aivazyian, I. S. Enyukov, and L. D. Meshalkin, *Applied statistics*, "Finansy i Statistika", Moscow, 1985. (Russian)
27. L. E. Sadovskiĭ, Bibl. "Kvant", 1, "Nauka", Moscow, 1974. (Russian)
28. I. I. Blekhman, A. D. Myshkis, and Ya. G. Panovko, *Mechanics and applied mathematics: The logic and special features of applications of mathematics*, "Nauka", Moscow, 1983. (Russian)
29. B. G. Litvak, *Expert information*, "Radio i svyaz", Moscow, 1982. (Russian)
30. L. E. Sadovskiĭ, A. L. Sadovskiĭ, and O. L. Sadovskaya, Bibl. "Kvant", 8, "Nauka", Moscow, 1984. (Russian)
31. A. E. Elo, *The rating of chessplayers—past and present*, London, 1978.
32. J. G. Kemeny and J. L. Snell, *Mathematical models in the social sciences*, MIT Press, Cambridge, MA, 1972.
33. Jessen Hoel, *Basic statistics for business and economics*, Wiley, New York, 1971.
34. S. I. Gass, *Linear programming. Methods and applications*, McGraw-Hill, New York, 1969.
35. M. H. Wagner, *Principles of operation research*, Prentice-Hall, New Jersey, 1969.
36. I.A.A.F/A.T.F.S., *Track & field statistics handbook for* 1984 *Los Angeles Olympic Games*, Los Angeles, CA, 1984.
37. V. Carter and R. E. Machol, *Operation research on football*, Oper. Res. **19** (1971), 541–544.

[28] This volume contains Russian translations of ten articles originally published in the Western literature, and of Abraham Wald's book *Statistical decision functions* (Wiley, New York; Chapman & Hall, London, 1950), plus seven original articles. (Editor's note)